The ABCs of Astrology and Astronomy

BOOKS BY

BERNICE PRILL GREBNER

Lunar Nodes

Decanates

Everything Has A Phase

Mercury, The Open Door, Part I

Mercury, The Open Door, Part II

American Federation of Astrologers

The Day Of Your Birth

Bee's Flight

The ABCs of Astrology and Astronomy

Grebner Books

The ABCs of Astrology and Astronomy

Bernice Prill Grebner

Grebner Books

Peoria Heights, Illinois

Grebner Books Publishing

5137 North Montclair Avenue

Peoria Heights, Illinois 61614-5221

First Printing 1993

ISBN 0-9626273-2-1

Printed in the United States of America

Graphics by Marjorie Grebner Welsch

Cover by H. Henry Welsch

This book is dedicated

to all people

who have an open mind

and strive to learn.

☆☆☆☆☆☆☆☆☆☆☆☆☆☆☆☆☆☆☆☆

INTRODUCTION

We must ever dedicate ourselves to reuniting and balancing mind, body and spirit, to reverse the centuries old separation of science (mind), body (physical) and spiritual (our soul and spiritual body). Religion said that astrology influences the physical, but argued that man was spiritual and not part of that. Then science came along and said we were not connected to it either. It all belongs together, body, mind and spirit; we must strive to understand we are one with the cosmic force. Everything in the solar system and the universe operates on physical, intellectual and spiritual principles.

February, 1993

Bernice Prill Grebner

PROLOGUE

Astrology can be complicated or it can be simple. All of this depends on one's awareness of the universe. If one is aware of the influence of the planets in our solar system and their effect upon each other, one begins to understand the whole system and its electromagnetic influence upon mankind.

The astrological chart (or map) is a person's guide to his place in the total scheme of things. If one follows one's own hunches and ideas, one will no doubt be following his system. If a person follows another's ideas instead of his own, he will be lost in time and space. Also, his health will diminish sooner than it should.

My grandfather, Ludwig Duhs, the first person in our family interested in astrology, argued "We would all be clones if each individual's personality and body was not determined by the planetary influence at the time of our birth. "That depends," he added, "on who or what establishes that system of influence."

Evangeline Adams, a great pioneer in astrology, said "Neptune answers the question 'Who is man?' Uranus 'Why is the man and What is his individual purpose?' And the inner planets show how he will do what he is designed to do and how he will come out."

As human beings we are constantly interacting with our neighbors, loved ones, and our environment. The trick is to still be ourselves without being controlled by these outer influences.

The early patriarchs themselves practised astrology and were talented in understanding its mysteries. Guido Bonatus, a famous Italian astrologer, said Jesus himself was well acquainted with its principles--as is revealed by His words. Cicero said the Babylonians perfected themselves in astrology by observing the birth maps of all the children for a period of forty seven, thousand years. I can't verify Cicero's report, but I believe it is probably correct.

Sir Francis Bacon said that astrology does show the natures and dispositions of men according to influences of the planets and those dispositions show different people are good at different things. Everyone can't be talented at the same thing. there has to be a larger pattern since all of us are pieces of the giant puzzle of life, not just on earth but as part of the cosmos.

The great astronomer Johannes Kepler wrote that an aspect is an angle formed at the earth of the beams of two planets or stars located at different distances, and that such aspects influence us.

Most of the astronomers of Europe and the known world during the middle ages delved into astrology. At that time astronomy and astrology were in harmony; they were one. You might say they were married. When there was disapproval, it came from the church.

The key to understanding astrology is to

associate key words with signs, aspects, planets, etc. Don't get in too deep at first. As you begin to understand the key words, you can add interpretation.

With this in mind, you should be ready for more information.

ASTROLOGY

ACTUAL SUN POSITION IN THE SIGNS

December 16 to January 18 - Sagittarius

January 19 to February 18 - Capricorn

February 19 to March 18 - Aquarius

March 19 to April 15 - Pisces

April 16 to May 15 - Aries

May 16 to June 15 - Taurus

June 16 to July 15 - Gemini

July 16 to August 15 - Cancer

August 16 to September 15 - Leo

September 16 to October 15 - Virgo

October 16 to November 15 - Libra

November 16 to December 15 - Scorpio

The Sun has moved out of the traditional signs into the preceding signs.

The modern horoscopes are not based on the actual position and aspects from the heavens necessarily but from a system that is internally self-consistent without regard to external positions or reality.

Perhaps we come to earth astrology. Compare yourself to the month you were born. If it is January, it is cold and enclosed. If it is March it is changeable and so are you. If it is hot summer, so is the weather and so is your own nature. As the Earth is influenced, so are we. We're not apart from this influence.

Maybe we could have just as easily called Capricorn, January; Aquarius, February; Pisces, March and so on. It is just a matter of semantics. Test it for yourself.

AIR SIGNS

Air controls the unifying aspects of life. Libra is concerned with deep relationships, including business and personal. Gemini is associated with brothers, sister, neighbors and community. Aquarius is associated with friendships and humanitarian causes. Air signs can be versatile and pleasing in personality. They enjoy interchange of ideas in companionship.

Although people controlled by air signs may seem superficial at times, they are not really so. They are interested in peace and intellectual growth above ordinary matter.

Air is the most erratic of the elements. Air is also the only element you can't see. Just because you can't see it, however, doesn't mean it doesn't have energy. Psychic energy, mental energy and stored physical energy are all in the air. One just needs to be in tune with it.

Air is the unifier of all around us. Without air we could not exist. Air balances.

Libra seeks to attract partners that unify lives.

Aquarius unifies by establishing bonds of mutual interest responsibility, whether in friendship or political causes.

Gemini communicates and unifies things in their environment and may also represent curiosity.

ASPECTS

The ancients considered squared aspects (90 degrees) and opposition (180 degrees) unfortunate and evil. By taking away the superstition and myth, all these aspects can be seen as energy meant to correct an unsolved problem in this lifetime. The energy at some point in the person's life is meant to be rechanneled into a productive area.

The so-called good aspects, trine (120 degrees) and sextile (60 degrees), represent problems that have been solved and achieved in a past lifetime. They represent where rewards are found in a present lifetime.

Fortunate Aspects (or productive talents)

Semi-sextile - 30 degrees

Sextile - 60 degrees

Quintile - 72 degrees

Trine - 120 degrees

Biquintile - 144 degrees

Unfortunate aspects (or unchanneled energy)

Semi-square - 45 degrees

Square - 90 degrees

Sesquiquadrate - 135 degrees

Opposition - 180 degrees

Overall consideration should include trines and sextiles with the productive aspects. Squares and oppostions for the unchanneled energy aspects should be considered at first examination. If a further detailed analysis is desired other minor aspects should be considered.

Each sign has 30 degrees. When calculating the Sun sign, every day is a degree. So the first day of a sign is one degree, the second day is two degrees and so on.

Psychological Meaning of the Aspects

Conjunction (orb of 8 degrees): emphasis, coming together

Semi-semiquintile (18 degrees, orb of 1

degree): appearances may be deceiving

Semi-sextile (30 degrees, orb of 3 degrees): difficulty in expression

Semi-quintile (36 degrees, orb of 1 degree): special techniques and talent

Nonagen (40 degrees, orb of 2 degrees): karmic, needs to break free

Semi-square (45 degrees, orb of 3 degrees): needs to do it themselves

Septile (51 degrees, orb of 1 degree): realizing reality, this is inevitable

Sextile (60 degrees, orb of 6 degrees): the ability of two forces working in cooperation, giving growth and success

Quintile (72 degrees, orb of 2 degrees): genius, spiritually connected to a working force of the universe

Square (90 degrees, orb of 8 degrees): excessive energy that has to be channeled, finding a way to make it work instead of having them work in a destructive clash. Squares are energy, like a car put into gear. One must get behind the wheel and direct it, or let it go off by itself and crash.

Trine (120 degrees, orb of 8 degrees): This is the capacity to receive cosmic strength without effort, something previously earned.

Sesquiquadrate (150 degrees, orb of 3 degrees): feeling of "aloneness in a crowd."

Bi-quintile (144 degrees, orb of 1 degree): ability to sublimate energy

Quincunx or inconjunct (180 degrees, orb of 1 degree): definite effort has to be made to arrive at success and fulfillment. This represents a definite influence. All other things will be just beyond reach until the right thing comes along.

Opposition (180 degrees, orb of 8 degrees): opposing views or interests; the need to see another's point of view; the need to achieve a balance (50-50). When that happens this aspect presents "both sides of the coin."

ASTEROIDS

The asteroids are between Mars and Jupiter. Some theorists think they are broken up planets: others think they originated as they are for a purpose.

There are four major asteroids:

Ceres is unconditional love and effort.

Pallas is cosmic wisdom.

Vesta is our area of dedication and commitment, and where this energy is either lowered or raised (like a pilot light). It can also represent sexual energy that is either controlled and redirected, or fulfilled in higher energy.

Juno is cosmic marriage and the ideal partner.

Minor asteroids:

Psyche is sensitivity to other people.

Eros is vitality and passion (lovers).

Lilith is the constructive release repressed anger.

Toro is the use and control power (warriors).

Sappho is romantic and artistic sensitivity.

Amor is spiritual or platonic love and compassion.

Pandora is curious and open to change.

Icarus is liberated and willing to take risks (freedom fighters).

Diana is good at survival and self-protection.

Hidalgo is assertive in defence of beliefs (protectors).

Urania is inspiring and knowledgeable.

Chiron is holistic understanding and breakthrough (holders of knowledge and breakthrough karma).

ASTRO-METEOROLOGY

Aries, if weather is disturbed, is hail and thunder.

First decanate: rainy and windy

Second decanate: temperate

Third decanate: hot and/or muggy

Taurus is temperate and warm.

First decanate: cloudy, windy, prone to earthquakes

Second decanate: moist, cool

Third decanate: stormy, with lightning and thunder

Gemini is temperate.

First decanate: moist and unpleasant

Second decanate: temperate

Third decanate: mixed and unpredictable

Cancer is gentle and warm.

First decanate: suffocating, heavy

Second decanate: temperate

Third decanate: windy

Leo is torrid.

First decanate: suffocating, prone to insect infestation

Second decanate: temperate

Third decanate: moist and unpleasant

Virgo is moist but thunderous.

First decanate: hot and unpleasant

Second decanate: temperate

Third decanate: moist and humid

Libra is variable.

First decanate: temperate

Second decanate: temperate

Third decanate: humid, rainy

Scorpio is thunderous and fiery.

First decanate: snowy

Second decanate: temperate

Third decanate: prone to earthquakes

Sagittarius is windy.

First decanate: moist

Second decanate: temperate

Third decanate: disturbed

Capricorn is very moist, rainy or snowy.

First decanate: unpleasant and cold

Second decanate: temperate

Third decanate: snowy, sleeting

Aquarius is cold and moist.

First decanate: moist

Second decanate: temperate

Third decanate: hot

Pisces is moist and windy.

First decanate: temperate

Second decanate: moist, rainy or snowy

Third decanate: hot

When Planets Are In Strong Aspect

Mercury: high winds and brisk conditions

Venus: pleasant weather with occasional showers

Mars: electric storms and turbulence

Jupiter: mild winters, warm summers, sultry conditions

Saturn: heavy rain or snow, gloomy weather

Uranus: chilly air, unpredictable weather

Neptune: winds, clouds

Pluto: heavy rain, storms

The Sun in relationship to the Moon: also the ruler of the phase of the Moon at the time; new Moon, crescent, first quarter, gibbous, full, disseminating, last quarter and balsamic.

New Moon: Leo (Sun)

Crescent: Cancer (Moon)

First Quarter: Sagittarius (Jupiter)

Gibbous: Capricorn (Saturn)

Full Moon: Virgo, Gemini (Mercury)

Disseminating: Taurus, Libra (Venus)

Last Quarter: Aquarius (Uranus)

Balsamic: Aries, Scorpio (Mars and Pluto)

ASTROLOGICAL SIGNS

Aries - March 21 to April 19 (I am)

Taurus - April 20 to May 20 (I have)

Gemini - May 21 to June 21 (I think)

Cancer - June 22 to July 22 (I feel)

Leo - July 23 to August 23 (I will)

Virgo - August 24 to September 22 (I serve or I criticize)

Libra - September 23 to October 23 (We are)

Scorpio - October 24 to November 22 (I am heroic)

Sagittarius - November 23 to December 22 (I understand)

Capricorn - December 23 to January 19 (I use)

Aquarius - January 20 to February 18 (I invent)

Pisces - February 19 to March 20 (I believe)

These beginning and ending dates of each sign may vary at times but they are close enough for most purposes.

Each person is like a planet influencing a given area. Among other things a planet's sphere of influence depends upon size and mass. This influence extends to us; to believe otherwise is to remain ignorant of the cosmic pattern of life.

♈ Aries: the ram, sheep or lamb

♉ Taurus: the bull, cow or ox

♊ Gemini: the twins, Castor and Pollux, or Pillars of Hercules

♋ Cancer: the crab, crawfish or scarab

♌ Leo: the lion and all members of the feline family

♍ Virgo: the virgin, a green branch, an ear of corn or wheat

♎ Libra: the balance, scales, either off center or at equilibrium

♏ Scorpio: the scorpion, eagle or snake

♐ Sagittarius: the archer, centaur or an arrow

♑ Capricorn: the sea gull, dolphin or scapegoat

♒ Aquarius: the water bearer, male, female or an androgynous figure pouring water from an urn to others

♓ Pisces: the fishes, the great fish or the fisherman

These key symbols should represent to each person studying astrology the temperaments of the signs and indicative of the qualities they have.

♈ The head and horns of a ram. The "hammer of the gods." The symbol of an offensive power.

♉ The head and horns of a bull. The sacred Apis, in whom the god Osiris was supposed to have been incarnate. A sepulcher.

♊ The figures of the twins, Castor and Pollux, as two pieces of wood bound together. It signifies the polarity of the mental process.

♋ The claws of the celestial crab. This symbol also represents the male and female seeds twisted together to make a whole.

♌ A corruption of the first letter of the Greek word for lion. Also the tail of the lion. the sign esoterically signifies the solar seed (the ego).

♍ The corruption of a Greek word signifying "virgin." This sign is supposed to represent the immaculate virgin, that is mental and spiritual virgin birth of self and others. It is not just a symbol of physical virginity.

♎ The symbol is a very crude effort to depict the balance needed not only for a relationship but also justice.

♏ The legs and tail of a scorpion, or simply the tail with the sting. The symbol is derived from the same figure as that of Virgo, but not as a virgin. It has deeper penetration here, whether sexually, intellectually or spiritually.

♐ An arrow with a short section of the bow. The symbol of enthusiasm and effort. Aiming at the stars or a higher and wider view of life.

♑ A corruption of the Greek word for goat. The goat with the fish's tail signifies extremes of height and success, the ability to climb to places others wouldn't dare attempt.

♒ Electromagnetic waves. Parallel lines of force that breaks through new ground for humanity.

♓ Two fish tied together at their tails with a length of rope. This is the symbol of bondage and captivity. It is either the inhibition of natural expression or sacrifice and dedication to a belief, an art. this symbol can express the intangible as one fish on the surface or repression of expression as the lower fish.

BRAIN FUNCTION RULERSHIP

The brain is divided into several parts and has rulership over certain physical functions, and is the master of certain impulses.

Moon: emotional faculties, hope, faith, sympathy, sensitivity, nurturing, emotionalism, among others. It is at the crown of the head.

Mercury: mental faculties, form, size, weight, color, order, number, experience, locality, time, tune, constructive talent, language, subjectivity. It is above and below the eyes.

Venus: physical faculies, love, caring, friendship, artistic talent, continuity of caring, the nature to dwell with another. It is at the base of the brain.

Sun: individuality, ego, personal sense of self. It is located between the eyes above the nose, or as some call it, the "third eye."

Mars: vitality, sexuality, action, combativeness, the drive to acquire things. It is located above the ears, behind the ears and close to the outside of the temples.

Jupiter: abstract ability, understanding, philosophical leanings, religious beliefs--if any, idealism, the percentage of optimistic spirit and higher objectivity. It is located in upper forehead.

Saturn: discipline, responsibility, dignity, firmness, justice, self esteem--or lack of, sense of authority in one's life. It is located at upper back of head.

Uranus does not govern any specific area of the brain, but can influence on a more comprehensive basis. It rules creative intellect, the ability for one to think for oneself, innovation and unconventionality--or lack of it.

Neptune also does not govern any specific area of the brain, but can influence on a more comprehensive basis. It rules psychic clairvoyance, music, dancing, spiritual faith and inspiration.

Pluto too does not govern and specific area of the brain but can influence on more comprehensive basis. It rules regeneration of body, mind, and spirit, and the ability to transform one's life. It rules healing, and when a person accepts healing energy.

The sign and house position of all the planets shows how these influences operate in a given lifetime.

CARDINAL SIGNS

To be a cardinal sign means being active!

Key words of the cardinal four are ambitious, enthusiastic, ardent, changeable, energetic, progressive, pioneering yet practical, useful, independent, proud, quick and insatiable. The negative traits are hasty, inconsiderate, reckless, destructive and domineering.

Aries--sometimes achieves goals through fearless force. They overcome obstacles by being aggressive and confident. You can usually find an Aries in a personal cause and in some kind of dispute.

Cancer--has silent, inconspicuous ambitions. They have a psychic instinct. Sometimes they get overemotional and possessive.

Libra--inclined to partnerships in business or personal affairs. They can be the diplomat, artist, musician or fashion designer.

Capricorn--excellent in business, ambitious, political, disciplined, responsible. They can be arrogant at times.

Houses:

First House--ruled by Aries, physical body and personality

Fourth House--ruled by Cancer, latter part of life and home

Seventh House--ruled by Libra, marriage and domestic conditions

Tenth House--ruled by Capricorn, public reputation and career

CHART CONSTRUCTION

Books and Material Needed

Logarithms chart with hours across top, minutes down. This helps you in determining the degree of the houses and ascendant with accuracy.

Table of houses book. This book shows the house cusp signs.

Longitude and latitude book of United States cities.

Longitude and latitude book of Foreign cities.

Time changes book. This is to determine what time changes have taken place.

Chart Form

Name

Birth date

Place (City)

Longitude and latitude

Time of day

Local time (variation from standard time) Each zone has a central point of time. Example: for Peoria, Illinois this amounts to 1 minute, 40 seconds.

Longitude correction . . . this is ten seconds for hours away from Greenwich Mean Time. Example: Peoria is 5 hours, 58 minutes, 20 seconds away from Greenwich . . . 59 second correction.

Elapsed Correction . . . time correction is 10 seconds per hour . . . Ephemeris books are calculated from midnight to midnight. Example: Peoria at 4:50 AM . . correction 48 seconds.

Sidereal Time . . . You will find this under the day considered in the ephemeris. Add the sidereal time to the total of local time, longitude correction and elapsed correction. That total will be the adjusted sidereal time. This is what you look for in the table of houses book, located at the top of each column. Now you are ready to construct your chart and put the signs and degrees on the houses.

Local Time--from above

G.M.T.-- Hours away from Greenwich

G.M.T. Adjusted-- Add the two figures above

Perm. Log. (Permanent Logarithm)-- Look on you Perm. Log. chart. Find hours on top, and minutes down. The point these meet is your permanent Logarithm number. This adjusts the degrees of the planets in their signs.

Check the ephemeris on the day you want. Look at the next day. Subtract the difference. Look on the Permanent log chart and see what the permanent log number is for the difference. Add this number to your log number. Find this on the perm. log chart and whatever degrees and minutes it denotes, add to the degrees in the ephemeris of the day in question. This gives you the accurate degrees at the time of day you were born. (The Table of Proportional Logarithms Chart is where you will find the log number).

Most ephemeris books before 1930 were calculated from noon to noon. After 1930 from midnight to midnight. Also, hours are based on 24-hour clocks. For example if you were born in 1929 at 2 PM, this is 2 hours. If you were born in 1937 at 2 AM this would be 2 hours.

It can get complicated, but not if you take one step at a time. If you have the right books, the right charts and understand what is happening, it all works out. Greenwich, England is where the central point is for calculating the astrological data for the daily ephemeris books so that is the reason we have to bring it to where the birth point is located. As in Peoria, Illinois we are five hours, 58 minutes, 20 seconds away from Greenwich. We add hours west of Greenwich. East of Greenwich is subtracted. You do not at this point need to know a lot of technical explanation for time and degree calculations. You just need to know what to do.

COLORS

The color representation of the astrological signs correspond to the colors of the rainbow. Red is on one side of the rainbow and violet on the other. The following colors for the signs seems to be the best.

Aries - brilliant red

Taurus - pink

Gemini - yellow

Cancer - white

Leo - orange

Virgo - indigo blue

Libra - medium blue

Scorpio - maroon

Sagittarius - royal blue

Capricorn - green

Aquarius - light sky blue

Pisces - violet

Each sign may rule many other colors, but

the ones listed above are the main ones.

CONCEPTION - Listed under PRE-NATAL EPOCH

CROSSES (Cardinal, Mutable, Fixed)

In regard to illness . . .

Crosses are either square to each other or opposite.

Cardinal--If one gets headaches ruled by Aries, these headaches may arise from stomach ruled by Cancer, kidneys ruled by Libra, or stiff joints ruled by Capricorn. The illnesses are interconnected. The thinking of Aries may result in illness described by Cancer, Libra or Capricorn.

Fixed--If there is a problem with a person's throat it may come from a Scorpio problem with the sexual or reproductive system. It can also come from a Leo heart problem or Aquarius nerves.

Mutable--These signs may find Feminine problems with the chest or breathing that can come from a Pisces foot problem or visa versa. Virgo intestinal problem can come from colds or visa versa. Sagittarius rules the liver and also rules appetite. Sometimes when the appetite gets out of line it causes liver problems, but could also cause related diseases ruled by the other mutable signs.

DAYS OF THE WEEK RULER

Moon rules Monday

Mars rules Tuesday

Mercury rules Wednesday

Jupiter rules Thursday

Venus rules Friday

Saturn rules Saturday

Sun rules Sunday

the ancients figured out this rulership. Whether it can be accurately proved or not there seems to be some influence operating here. Prove it for yourself. Its value comes in how it works or doesn't for the individual.

DECANATES

Each zodiacal sign is divided into three-ten degree sections so that each sign has three decanate divisions.

Each day of a sign represents one degree. Thus, the first ten days of a sign usually represent the first decanate of that sign, the second ten days the second decanate of the sign and the last ten days of a sign the third and final decanate.

The first decanate influences the physical nature, the second decanate the mental nature and the third decanate the spiritual.

FIRST DECANATE

Aries--Aries: March 21 - 30, the pioneer

(Mars--Mars)

SECOND DECANATE

Aries--Leo: March 30 - April 10, the gambler

(Mars--Sun)

THIRD DECANATE

Aries--Sagittarius: April 11 - April 20, the hero

(Mars--Jupiter)

FIRST DECANATE

Taurus--Taurus: April 21 - April 30, the singer

(Venus--Venus)

SECOND DECANATE

Taurus--Virgo: May 1 - May 10, the gardener

(Venus-Mercury)

THIRD DECANATE

Taurus--Capricorn: May 11 - May 21, the banker

(Venus--Saturn)

FIRST DECANATE

Gemini--Gemini: May 22 - May 31, the writer

(Mercury--Mercury)

SECOND DECANATE

Gemini--Libra: June 1 - June 11, the diplomat

(Mercury--Venus)

THIRD DECANATE

Gemini--Aquarius: June 12 - June 21, the lecturer

(Mercury--Uranus)

FIRST DECANATE

Cancer--Cancer: June 22 - July 2, the historian

(Moon--Moon)

SECOND DECANATE

Cancer--Scorpio: July 3 - July 12, the prophet

(Moon-Pluto)

THIRD DECANATE

Cancer--Pisces: July 13 - July 22, the musician

(Cancer--Neptune)

FIRST DECANATE

Leo--Leo: July 23 - August 2, the actor

(Sun--Sun)

SECOND DECANATE

Leo--Sagittarius: August 3 - August 12, the promoter

(Sun--Jupiter)

THIRD DECANATE

Leo--Aries: August 13 - August 23, the explorer

(Sun--Mars)

FIRST DECANATE

Virgo--Virgo: August 24 - September 2, the analyst

(Mercury-Mercury)

Note: Some experts think newly discovered Chiron rules Virgo.

SECOND DECANATE

Virgo--Capricorn: September 3 - September 12, the teacher

(Mercury--Saturn)

THIRD DECANATE

Virgo--Taurus: September 13 - September 23, the precisionist

(Mercury--Venus)

FIRST DECANATE

Libra--Libra: September 24 - October 3, the artist

(Venus--Venus)

SECOND DECANATE

Libra--Aquarius: October 4 - October 13, the debater

(Venus--Uranus)

THIRD DECANATE

Libra--Gemini: October 14 - October 23, the reporter

(Venus--Mercury)

FIRST DECANATE

Scorpio--Scorpio: October 24 - November 2, the magician

(Pluto--Pluto)

SECOND DECANATE

Scorpio--Pisces: November 3 - November 12, the researcher

(Pluto--Neptune)

THIRD DECANATE

Scorpio--Cancer: November 13 - November 22, the detective

(Pluto--Moon)

FIRST DECANATE

Sagittarius--Sagittarius: November 23 - December 2, the evangelist

(Jupiter--Jupiter)

SECOND DECANATE

Sagittarius--Aries: December 3 - December 12, the philosopher

(Jupiter--Mars)

THIRD DECANATE

Sagittarius--Leo: December 13 - December 22, the ambassador

(Jupiter-Sun)

FIRST DECANATE

Capricorn--Capricorn: December 22 - December 31, the executive

(Saturn--Saturn)

SECOND DECANATE

Capricorn--Taurus: January 1 - January 10, the architect

(Saturn--Venus)

THIRD DECANATE

Capricorn--Virgo: January 11 - January 20, the idealist

(Saturn--Mercury)

FIRST DECANATE

Aquarius--Aquarius: January 21 - January 30, the scientist

(Uranus--Uranus)

SECOND DECANATE

Aquarius--Gemini: January 31 - February 9, the pilot

(Uranus--Mercury)

THIRD DECANATE

Aquarius--Libra: February 10 - February 19, the humanitarian

(Uranus--Venus)

FIRST DECANATE

Pisces--Pisces: February 20 - March 1, the poet/musician

(Neptune--Neptune)

SECOND DECANATE

Pisces--Cancer: March 2 - March 11, the dreamer

(Neptune--Moon)

THIRD DECANATE

Pisces--Scorpio: March 12 - March 21, the mystic

(Neptune--Pluto)

DETERMINING SEX OF A BABY

The sex of a baby may be determined by

the power of the current positions of the planets, signs and houses. It is as simple as determining if a majority of elements are masculine signs, the baby will be male; if a majority of elements are in feminine signs, the baby will be female.

Probability of twins can be determined if conception came when the Sun was in a dual sign; the Moon in a dual sign; the ascendant is also in a dual sign and a dual sign on the fifth house. Chances for all these to take place is rare as are twins.

Majority of planets in Sagittarius may lengthen the upper legs of the baby. Aquarius may increase the size of the ankles. Taurus may increase the size of the neck. Gemini may lengthen or shorten the arms. Controlling planets in Cancer may cause stoutness, while an excessively strong Capricorn may cause premature aging. Aries may cause early baldness and/or freckles.

DISEASES

Astrological Signs

Aries--headaches, epilepsy, fevers, neuralgia, congestion, eye problems, eruptions on face and head, inflammation, insomnia, serious diseases in general, apoplexy, wounds and accidents

Taurus--diphtheria, goiter, glandular swellings in neck, strangulation, quinsy, tonsillitis, gangrene, impure blood, obesity, impaired circulation, diabetes and vein constriction

Gemini--bronchitis, pulmonary consumption, nervous diseases, pneumonia, pleurisy, fractures of shoulders, arms and hands, often chronic ailments, asthma and failing of sense perceptions

Cancer--Digestive ailments, dropsy, coughs, cancer, tumors, ovarian trouble, bronchitis, flatulency, stomach problems, sclerosis and sometimes delicate constitution

Leo--heart problems, especially angina

pectoris, palpitations, poor circulation, fevers, spinal problems, fainting spells

Virgo--afflictions of the intestines and digestive tract, cholera, peritonitis, malnutrition, typhoid, tapeworm, appendicitis, possibility of general paralysis in old age

Libra--inflammation of kidneys, Bright's disease, all forms of kidney problems, lumbago, diabetes, easily thrown off balance in health but quickly responds to treatment

Scorpio--ruptures, hemorrhoids, diseases of the generative system, any type of sexual disease, ulcers, toxic complaints, nasal catarrh, prostate problems

Sagittarius--rheumatism, coxalgia, dislocation of hip joints, gout, paralysis, sciatica, accident prone and liver complaints

Capricorn--skin diseases, articular rheumatism, gout, hysteria, cramps, infantile paralysis, deformities of the bone structure, broken bones and/or dislocation of bones and toothaches

Aquarius--sprained and broken ankles, anemia, swollen ankles, cramps, spasmodic and nervous diseases, blood poisoning, varicose veins and sensitive skin

Pisces--colds, deformities of feet and toes, disease of mucous discharges and secretions, tumors, bunions, alcoholism, sexual diseases, tuberculosis and liver problems

These above diseases would be indicated by the person's ascendant sign and the sign on the cusp of the sixth house, and Moon sign, not just the Sun sign. If you know these three signs you can better prepare yourself for a weakness in this area and do something about it. Disease can be avoided with awareness.

Planets/Signs

The sign of the ascendant is the first house and the sign on the cusp of the sixth house of the astrological chart determine disease tendencies and influences. Sometimes the Moon sign can chow susceptibility to disease. The ascendant is the time of day a person is born.

Moon/Cancer--dropsy, tumors, abscesses, stomach trouble, female disorders, problems with mucous membranes, epilepsy and periodic complaints brought on by excess emotionalism

Mercury/Gemini, Virgo--nervous disorders, headaches due to stress, respiratory difficulties, delirium, neurasthenia and speech problems

Venus/Libra, Taurus--throat infections, goiter, lymphatic ailments and diabetes

Sun/Leo--problems with heart and arteries, spinal ailments, sunstroke, eye disorders, fainting spells, fevers and diseases of the spleen

Mars/Aries--inflammations and acute fevers, infections, muscular and genital disorders, hemorrhaging and infectious and contagious diseases in general

Jupiter/Sagittarius--blood disorder, liver trouble, pleurisy, fatty deposits, diabetes and high blood pressure

Saturn/Capricorn--chronic diseases, rheumatism, arthritis, gout, paralysis, skin diseases, gangrene, impeded circulation, hardening of the arteries and atrophy

Uranus/Aquarius--spasmodic disorders, acute nervous conditions, ruptures, cramps, spasms, malnutrition and circulation problems

Neptune/Pisces--obscure and difficult to determine diseases, poisoning, drug habits, psychic disturbances, hallucinations and hypochondria

Pluto/Scorpio--sexual diseases, tumors and problems in childbearing

EARTH ASTROLOGY

Some Native American Indians followed these divisions of the year.

December 22 - January 19: Earth Renewal (Capricorn)

January 10 - February 18: Rest and Cleansing (Aquarius)

February 19 - March 20: Big Winds (Pisces)

March 21 - April 19: Budding Trees (Aries)

April 20 - May 20: Frogs Return (Taurus)

May 21 - June 20: Corn Planting (Gemini)

June 21 - July 22: Strong Sun (Cancer)

July 23 - August 22: Ripe Berries (Leo)

August 23 - September 22: Harvest (Virgo)

September 23 - October 23: Ducks Fly (Libra)

October 24 - November 21: Freeze Up (Scorpio)

November 22 - December 21: Long Snows (Sagittarius)

As the seasons pass and return our EARTH is influenced by its association with a greater system; so are we.

EARTH SIGNS

Earth signs rule materialistic, physical areas.

Earth triplicity controls the environment. Capricorn controls social standing and the person's place in his world. Virgo rules health conditions and the working environment. Taurus rules the resources and financial conditions.

Earth signs show good recuperative power, conservation and good control of resources.

People dominated by earth signs are inhibited in self expression and so must work on this. Their temperament is more passive. They are also lovers of material comfort and will make every effort to have it.

Capricorn represents the rugged peaks of the earth, mountains. Capricorn people usually impress others with a sense of majesty.

Virgo represents the deep valley where

secrets lie hidden and that means Virgo people have a lot more on the ball than others think. Sometimes they feel the weight of the world on their shoulders. At their best they want to serve and do a good job.

Taurus symbolizes the fertile plains, and broad farm lands ready to be plowed by the great ox that rules them. They are warm and stubborn but they do produce things that easily turn into money or food for others.

EMOTIONAL NATURE

Cancer/Moon--kindness, sympathy, modesty, love of animals, sense of wanting to nurture, maternal instincts, shy, sensitive to hurt

Mercury/Gemini-Virgo--analytical thinker who strives to acquire knowledge because it gives emotional satisfaction, happiness in being proficient, skillful with words, good at debate, emotional satisfaction at expression in writing

Venus/Libra-Taurus--emotional satisfaction in love attachments stimulated by beauty, art, environment, or a person, not analytical, responsive to others, caring

Sun/Leo--dignified in emotions dramatic, showy, patronizing at times, but always ascending to dignity

Mars/Aries--enthusiastic in emotional expression, ardent, pushes for satisfaction of emotional desires even when discouraged, original in expression, dares to be and to do

Jupiter/Sagittarius--blending mind with heart, feelings and emotions, thoughts tempered with mercy and understanding, optimistic

Saturn/Capricorn--reserved, inhibited at times, self-controlled, just, patient and prudent

Uranus/Aquarius--original, inventive, emotionally spontaneous, in love with liberty, idealistic toward humanity's betterment, always wants to establish a new order or things

Neptune/Pisces--idealistic, romantic, musical, psychic, feels things other do not even realize exists, tolerant, compassionate

Pluto/Scorpio--intense, dominating, secretive, heroic and able to rise above any and all difficulties standing in the way of emotional fulfillment

EMOTIONAL TENDENCIES

Aries--courageous, enthusiastic, energetic, proud, impulsive,quick, intemperate

Taurus--amorous, artistic, loyal, gentle, domestic, quick-tempered, stubborn, self-indulgent, sensual

Gemini-eloquent, humane, fond of travel, friendly, changeable, ungrateful at times, easily distracted, emotionally expressive

Cancer--dreamy, maternal, kind-hearted, romantic, domestic, impressionable, psychic, some times despondent, sometimes lazy and self-absorbed

Leo--faithful, affectionate, idealistic, proud, chivalrous, excitable, vain, occasionally cruel and selfish, sometimes generous and merciful

Virgo--kindly, highly evolved, altruistic, fussy, petty, sometimes superficial in love and emotions

Scorpio--capable of extremes in emotion; all or nothing; when highly evolved--impersonal, unselfish, imaginative; when not highly evolved--revengeful, secretive, quick-tempered, suspicious, and either heroic or cowardly

Sagittarius--emotionally impersonal, proud, zealous, energetic, friendly, openhearted, understanding, idealistic, sincere, impatient, daring, not domestic, speculative

Capricorn--emotionally inhibited, sometimes self-pitying, unforgiving, cold (or seemingly so), irritable, mentally directed (as opposed to emotionally)

Aquarius--emotional, very active, vivacious, excitable, kindly, well-liked, gentle, altruistic, domestic but changeable, unconventional, worrying

Pisces--emotionally inhibited, sensitive and impressionable, psychic, somewhat self-sacrificing, melancholy, misunderstood, not very assertive

FIRE SIGNS

Fire signs control the expressions of creativity. Aries is concerned with building the body of the person and his individual uniqueness. Leo deals with offspring whether those offspring are physical children, mental or spiritual children (such as creative writing). Sagittarius rules the development of the higher mind, philosophy, religion and abstract sciences.

Fire signs have fantastic energy and seem to be continually active. They are impulsive, sharp, brilliant, original and talented in teaching others science and philosophy. You name it, they can teach it. Their temperament is dynamic, and sometimes forceful.

Fire signs continually want to burn away the dross and liberate the spirit to higher values.

Aries is fire that is red and angry. There is great heat here, and the smoldering flames can burst out again at any time. When dealing with an Aries, don't underestimate them. This is fire for self.

Leo's flame is golden orange and burns in a steady glow, warming the world around it. It is for all to see and be influenced by.

Sagittarius is a fire that inspires others with its clear, blue flame. It inspires them to be all they can be mentally, by stimulating the intellectual recesses of the mind.

FIXED SIGNS

The fixed signs are dogmatic, absorbed, determined, accumulative, firm, conventional or unconventional depending upon the situation involved, proud, thrifty, subject to habit and unchanging attitudes and life-styles, bossy, dignified and constructive.

These people's minds are penetrating, analytical, secretive, good at remembering, slow but sure.

Undeveloped fixed signs are stubborn, bigoted, sluggish, egotistical, tyrannical and too concerned with their way.

Taurus is stubborn with his opinions almost to the point of closed-mindedness, inclined to be jealous, prudent, tenacious, unwilling to admit defeat.

Leo is willing to suffer for a cause, a principle or a loved one, determined in the face of obstacles or opposition, pompous, dramatic, entertaining

Scorpio recognizes no authority as equal to his experience and his fixed opinion. Yet he does not let his opinions interfere with the rights of others. He is enigmatic: at times saintly, at times devilish.

Aquarius can be unconventional, intellectually creative and committed to humanitarian movements. Just when he seems unchangeable he'll take sudden action that surprises everyone. He can accomplish things and win over people without becoming unpopular. The Aquarian can do things others would not even dream of doing.

Houses:

Second House--personal financial influences

Fifth House--one's children, pleasures, creative expression

Eighth House--causes of death, regeneration, benefits of partnership (personal or business)

Eleventh House--attitudes with friends, goals and objectives, humanitarian causes (if any)

GEOCENTRIC ASTROLOGY

Geocentric is earth centered, and the

positions of the planets in the signs are so calculated. Because we are on earth and not on the Sun (we would not last long there) we are going to follow the geocentric chart; this chart indicates purposes on earth.

Heliocentric astrology is calculated from the Sun and indicates planets in the signs from that position. This changes the inner planets in the signs. It seems to indicate our higher self in relationship to a higher purpose, or our soul position in the universe.

The Ancients thought the Earth was the center of the universe and so the calculations were from Earth and were studied from that position. Nicolaus Copernicus proved the Earth was not the central position in our solar system but that the Sun was. This was contrary to the opinion at the time. His system was fully worked out by 1515 but it was not until 1543 that it was published. At that time the Catholic Church denounced his findings.

Heliocentric astrology is Sun centered; the planets--including Earth--revolve about the Sun and the Moon revolves around the Earth. Astrology follows the ancient system of calculation called geocentric, which holds the Earth to be the center of our solar system and the Sun as the fourth planet figuring from our Earth. That is the way astrologers view it because we are on the Earth and the influences come to us on the Earth. We are not on the Sun; that is what is presumed to make the difference in influence. Although the geocentric system is not astronomically correct, it has proved itself to be astrologically true over a long period of time in its effect on man and Earth. As the person or place whose chart is being considered is upon the Earth, and that makes the Earth for him the center of the solar system.

Recently some astrologers have made an effort to determine heliocentric influences but with limited success.

Geocentric astrology is concerned with material things, physical affairs on Earth.

Heliocentric astrology is concerned with spiritual levels of the individual. We do not yet know enough this, however, to accurately calculate heliocentric influences.

Incidentally, according to modern scientific calculations our Earth is about four and a half billion years old. Life began about three and a half billion years ago. "Human" life dates back several million years.

HELIOCENTRIC ASTROLOGY - See Geocentric Astrology

HORARY ASTROLOGY

When someone asks a question, a chart is set up for the moment the question is asked. Horary astrology is used to find the answer.

The Moon and planets signify certain people, places and conditions.

Moon--sailors, servants, women, mother, home, food, restaurants

Mercury--merchants, sisters and brothers, community, environment, communication, letters clever people; negative influence--thieves, people who lie or deceive

Venus--lovers, musicians, artists, clothiers, interior or exterior decorators, young women, perfume, cosmetics, money or any kind of desire

Mars--soldiers, men, sex, any kind of action, sports, mechanics, aggressive people; negative influence--uncultured people, anger, revengeful people

Sun--main man in a woman's life, educator, powerful person, honor, ego, nobility; negative influence-- a man who is egotistical

Jupiter--professional person, religious person, judge, lawyer, counsellor, a person in his 50's, scientist, foreigner, rich and generous friend, a person with integrity; negative influence--black sheep of the family, person who exaggerates or promises more than he can deliver

Saturn--aged person over 70, conservative person, father, person in authority, laborer, business leader; negative influence--beggars, clowns, paupers, recluses, narrow and prudent people, excessively traditional people

Uranus--unexpected situations, inventors, electricians, innovative people, important proposition or opportunity, change, astrologer, humanitarian, friend, mentally creative people; negative influence--loss from unwise investments, impostors, erratic people

Neptune--psychic person, metaphysics, magicians, mystics, wise and eccentric people, anyone engaged in water, ocean or river pursuits, hypnotists, musical-poetic geniuses, prophets, spiritual counsellors, an sacrificial person, a compassionate person; negative influence--and illusory situation, deception, a self-pitying person

Pluto--surgeon, doctor, hero, researcher, death, transformation, regeneration; negative influence--domineering person, cruelty, extremes

The twelve houses:

Horary astrology is asking a question at a given point in time, setting up a chart for that moment, and getting the answers from that chart.

First House--all questions concerning the life and well being of the person asking the questions, the effect of various actions on his life, accidents, injuries

Second House--losses or gains, outcome of financial transactions, loans, movable property, personal belongings

Third house--changes and moves in business or personal enterprises, neighbors, relatives, brothers and sisters, short trips, documents to be signed, messages, news from others, letters, etc.

Fourth House--the person's nurturing parent (could be father or mother but is usually the mother), land, houses, estates, town, end of life, treasures within a home

Fifth House--pregnancy, property belonging to the fourth house parent, children, number of children, gambling, fun and games, amusement places

Sixth House--servants, business employees, those who are dependent upon the person, smaller animals, farmers, health of the person, relatives on the mother's side, ability to serve others

Seventh House--the part the astrologer plays in the investigation of the answers for the person, marriage, where the person may move to if he changes his living quarters, business partnerships, dealing with law breakers

Eighth House--property of the person's partner, but also property they may have in common, manner of death, dispositions of the person's property, legacies, wills, money from partner or from family wills, sexual nature, procreation tendencies

Ninth House--marriage partner's relatives, meaning of dreams and visions, the accumulation of knowledge in higher educational institutions, religious life of the person, changes in religious matters, foreigners or long distant trips

Tenth House--careers, father, public recognition, securing money or pensions from the government, authority influence in the person's life other than father

Eleventh House--friendships, goals, objectives, social connections--whether helpful or disruptive, desires and wishes, political views, causes the person may get into

Twelfth House--imprisonment (whether institutional or personal), persecutions, secret relationships, secret loves or secret enemies, secret sorrows and joys, psychic influence the person may have, musical or poetic inspirations, subconscious directives (good and bad)

HOUSE MEANINGS

First House--personality people first see,

and your physical body

Second House--one's resources (income), also talent for doing well in this life

Third House--one's environmental influence, give and take, neighbors, brothers and sisters and community influence

Fourth House--one parent (the nurturing parent) domestic influence, the heart of the person, the person most people don't know until they live with him

Fifth House--one's children (biological or creative), gambling, how the person has fun

Sixth House--health and working conditions (good conditions-good health; poor conditions-poor health), how the person serves

Seventh House--marriage partners, any deep relationships, business partners

Eighth House--inheritance, what the person expects from a partnership, how he rejuvenates himself

Ninth House--higher mind, where one can be objective, religion, philosophy, distant places, relatives of the marriage partner, science

Tenth House--career, public image, father figure influence, views

Eleventh House--friends, goals and objectives, causes, political views

Twelfth House--things hidden behind the scenes, where sacrifices can be made, influences from a past life that may or may not be subconscious, karmic influences the person has to work out

The sign on each of these houses and the planetary influence in each one tells the story.

HOUSE RULERSHIP

First House--Aries

Second House--Taurus

Third House--Gemini

Fourth House--Cancer

Fifth House--Leo

Sixth House--Virgo

Seventh House--Libra

Eighth House--Scorpio

Ninth House--Sagittarius

Tenth House--Capricorn

Eleventh House--Aquarius

Twelfth House--Pisces

JEWELS

For a little history of the twelve astrological sign jewels we must go back to Jewish history. There were twelve tribes of Israel, which according to symbolism, are the twelve zodiacal constellations. Josephus declared that the jewels by their color and light reveal their spiritual strength.

Opinions vary as to which stones should be assigned to various signs.

Aries---diamond

Taurus--emerald

Gemini--aquamarine

Cancer--pearl

Leo--ruby

Aquarius--amethyst

Virgo--topaz

Libra--coral

Scorpio--bloodstone

Sagittarius--turquoise

Capricorn--onyx

Pisces--opal

KEPLER, *Johannes* (1571-1630)

Kepler, born on December 27, 1571, was the founder of modern astronomy and was also an astrologer. He was criticized by the scientific community for his interest in astrology. But it was through astrology that he earned his livelihood. He was delicate in health. Smallpox contracted at the age of four left him with a partial infirmity of the hands and eyes.

He was poor all his life and neglected in comparison to his brothers and sisters. He completed his studies and was appointed head of astronomy and mathematics at Graz in 1594.

Kepler made concessions to astrology and was criticized. He answered one critic with the words, "I am not going to argue with you, I can see you have not studied it, I have."

Kepler had to win his mother's release from prison where she was placed in 1620 under a charge of commerce with the devil, and was almost put to death as a witch. He also said, "Astrology can be a very foolish child; but heavens! where would astronomy, which is so wise, be without her crazy child?" (In other words they need each other if they could but see that.)

Kepler was a genius--which is shown by his Mercury quintile Jupiter. His instincts and intuition were great. He established Kepler's Laws, the first of which states:

the orbit of every planet is an ellipse.

Kepler had Sun in Capricorn, Moon in Gemini, Mercury in Capricorn, Venus in Capricorn, Jupiter in Pisces and Saturn in Scorpio, Mars in Libra. His Sun and Venus were Sextile Jupiter and Saturn. He had understanding and could discipline himself to whatever person and project he set his heart and mind to do. His Moon was trine Mars, which means his emotional nature was curious and interested in the mysteries of the environment and the universe. This Moon trined Mars means his energy and passion were in complete harmony with his emotions and feelings. Once he felt something he took action with it. Even though his health was poor he accomplished more work with his Mercury square Mars than other people with higher energy. Jupiter trines Saturn provided for success in the world and an understanding of what he could give the world.

This is a short description of the astronomer who was also an astrologer. Kepler knew the two disciplines needed each other. He was a great man. If only more had been more like him. He could take the intangible and make it real for others.

LIVES (past, present and future)

The sign of the ascendant becomes the present. This is now. The sign on the cusp of the twelfth can indicate a past life. The sign on the cusp of the second becomes a future ascendant.

The first, fourth, seventh and tenth are supposed to deal with the present--the angular houses. The succeedant houses--second, fifth, eighth and eleventh--become symbolic of the future task the person faces in his evolutionary growth through incarnation. The cadent houses--third, sixth, ninth, and twelfth--reveal past conditions, the state of the ego in a previous existence. This, then, shows the spiritual state of the person.

LUNAR ECLIPSE

A lunar eclipse is when the Earth is between the Sun and the Moon and the Earth casts a shadow on the Moon.

Lunar eclipses (unlike solar eclipses) are almost immediate in their effect, and correspondingly of shorter duration. The solar eclipse is considered to be influential for many years since the eclipse sometimes lasts hours (a year for each). Lunar eclipses are influential at least a year.

Let me quote something from *The Bible.*

The 21st Chapter of Luke . . .

"There will be signs in the sun and moon and stars; on earth nations in agony, bewildered by the clamour of the ocean and its waves . . . . " (*The Jerusalem Bible*, Luke 21:25)

LUNAR NODES

The point in the orbit of the moon where it crosses the ecliptic from south to north is called the north (ascending) node. It's symbol is ☊ and it was known to the ancients as Caput Draconis, or the Dragon's Head. The point in the orbit of the Moon as it crosses the ecliptic from north to south is the south (descending) node. It's symbol is ☋ and it was known to the ancients as Cauda Draconis or the Dragon's Tail. All the planets other than the Sun have their nodes, but astrology concerns itself with the Moon's nodes because of its proximity to Earth. the head of the dragon is always 180 degrees from its tail.

These are hypothetical points. While the Moon's nodes, like the Part of Fortune (also a hypothetical position), have been the subject of controversy in astrology, your author believes they work.

Refer to my book *Lunar Nodes New Concepts*, published by The American Federation of Astrologers in Tempe, Arizona. This book will also give you an ephemeris of the lunar nodes through the signs by month, day and year.

The north node is regarded as benefic or the place for new growth in this lifetime (reincarnation). It is supposed to represent what we have to learn. It is ruled by Jupiter and Venus. the south node is regarded by some people as more malefic. But it is simply what we already know from another life, or what we were born with as a natural talent. But it is also where we have to give to others without return. It is ruled by Saturn and Mars. The North Node is where energy, light and benefits come to us. The influence of the lunar nodes have a small orb of influence, three degrees.

South Node--Aries (first house)--egotism

South Node--Taurus (second house)--determination

South Node--Gemini (third house)--subjectivity

South Node--Cancer (fourth house)--introversion

South Node--Leo (fifth house)--love or acting

South Node--Virgo (sixth house)--criticism or work

South Node--Libra (seventh house)--indulgence with partnerships

South Node--Scorpio (eighth house)--secretiveness

South Node--Sagittarius (ninth house)--objectivity

South Node--Capricorn (tenth house)extroversion or executive tendencies

South Node--Aquarius (eleventh house)--unattached love

South Node--Pisces (twelfth house)--psychic sensitivity

Since the thing of the north node position are the things one has to learn, it represents the learning of the sign and house position in which they are located. For example: Aries--learning to do for himself; Taurus--learning to make his own resources; Gemini--learning to express himself and deal with environment; Cancer--learning to nurture family and home; Leo--learning one to one love; Virgo--learning to work and be of service to others and learning the art of discrimination; Libra--learning to relate to a partner; Scorpio--learning to accept a partner's help or the true beauty of sexual contacts; Sagittarius--learning objectivity and higher mind subjects; Capricorn--vocational success and learning to deal with a father or authority figure; Aquarius--learning friendship and detachment in love; Pisces--self sacrifice and compassion.

MALEFIC DESCRIPTIONS

Violent sign--Scorpio

Fruitful sign--Cancer

Barren sign--Aries

Commanding Signs--Aries, Sagittarius

Vocal signs--Gemini, Aquarius

Mute signs--Pisces, Scorpio

Imperfect signs--Pisces, Capricorn

Strong signs--Scorpio, Aquarius

Weak signs--Cancer, Capricorn, Pisces

MENTAL CHARACTERISTICS

Moon/Cancer--instinctual, imaginative, family oriented, receptive, intuitive, psychic, fluid, easily taken off guard

Mercury/Virgo-Gemini--expressive, rational, thoughtful, eloquent, good memorization, detail-oriented, good at debating

Venus/Libra-Taurus--socially oriented, artistic, dedicated to loved ones, imaginative, affable

Sun/Leo--good at leadership, drama, strong-minded, generous, honest, creative, strong egoism

Mars/Aries--innovative, active, courageous in new ideas, enthusiastic, self-assured

Jupiter/Sagittarius--sensible and understanding, humane, optimistic, perceptive, objective, open-minded

Saturn/Capricorn--detail-oriented, accurate, persevering, organized, sensible, practical

Uranus/Aquarius--independent in thought, original, uncontrollable, inventive, altruistic, flexible in attitude, very perceptive of others motives

Neptune/Pisces--able to conceive things inexplicable to others, psychic, mystical, musical, successful in things where the person has to tune in on the intangibles beyond the average mind

Pluto/Scorpio--research-oriented, penetrating, good at detective work, heroic and thorough, sometimes dominating

METALS

Moon, Cancer--silver, aluminum

Mercury, Gemini--quicksilver (mercury)

Venus, Libra--copper, tungsten

Sun, Leo--gold

Mars, Aries--steel

Jupiter, Sagittarius--tin

Saturn, Capricorn--lead

Uranus, Aquarius--uranium, radium

Neptune, Pisces--platinum

Pluto, Scorpio--iron

Earth, Taurus--brass

Chiron, Virgo--zinc

MINERAL QUALITIES

Earth, Taurus--brittle substances

Moon, Cancer--smooth, soft substances

Mercury, Gemini--flowing and veined substances

Venus, Libra--substances that have high polish or brilliance

Sun, Leo--substances that are very luminous

Mars, Aries--sparkling and/or fiery substances

Jupiter, Sagittarius--substances that expand easily

Saturn, Capricorn--dull and heavy substances

Uranus, Aquarius--electrically charged substances

Neptune, Pisces--substances that are dark and not easily identified

Pluto, Scorpio--healing substances, transforming substances, substances that change their nature

Chiron, Virgo--substances that are small, but very strong

MISDIRECTED ENERGY

Aries--Excess of energy can cause angry outbursts. If this happens too often, diseases of Aries takes place.

Taurus--Brooding, stubborn excess, too rich a diet and too much of it, too much attention to material things and luxury leads to Taurus diseases.

Gemini--Restlessness, scattered attention, detrimental environmental conditions which keep the person in constant tension can bring on the diseases of Gemini.

Cancer--Distrust of other people, grudges or dislike of others

and greed will lead to Cancer diseases.

Leo--Impulsiveness, egotistical tendencies, desire to accomplish things at all costs, and sometimes excessive ego will cause Leo diseases.

Virgo--Selfishness and self pity, critical attitudes, perfectionism may cause diseases of Virgo, certainly an acid condition that depletes one's energy.

Libra--Melancholy spells, jealously, and constant indecision will prematurely age the person and may lead to other Libra diseases.

Scorpio--Tyrannical tendencies, destructiveness, excessive nostalgia and domination of others will lead to diseases of Scorpio one way or another and shorten the person's life.

Sagittarius--Injury-producing adventurousness, restlessness, and a know-it-all type of attitude (without listening) leads to Sagittarian diseases.

Capricorn--Crystallization of attitudes toward life, too much self-centeredness (my way and not yours), and closed-mindedness leads to Capricorn diseases.

Aquarius--Supersensitivity (even though the person usually tries to forget himself through action) and fidgety efforts lead to Aquarian diseases!

Pisces--Sensitivity, secretiveness, a wild imagination and out of touch with reality lead to imbalance and Pisces diseases.

MUNDANE ASTROLOGY

Mundane astrology concerns the influence of the planets and the zodiacal signs upon countries, cities and various functions of Earth life.

The houses show the divisions of the social order.

First House--general conditions of a country, the ordinary populace, public health and well being, the general character of the people living in the country

Second House--the country's finances, the revenue of the country, stocks and bonds, banks, commercial trade, and possible fluctuations of the value of money

Third House--telephone systems, the postal system, traffic, motor cars, books and newspapers, immediate surrounding countries, and method of transportation

Fourth House--agriculture, weather, crops, disputes over territory and land, public buildings, in politics the party opposing the government in power, people against authority, family life of the people

Fifth House--the birth rate, theatres and

places of entertainment, educational systems, public morale, speculation and gambling, social functions, diplomats

Sixth House--army, navy, marines, warships, public health, civil service, working class people, strikes

Seventh House--international relations,arbitration, war, marriages of the people, divorces, foreign trade, general conditions of relationships between men and women and how they treat each other

Eighth House--the death rate, suicides, private councils, financial relations with foreign countries, funerals, results of partnerships (business and personal)

Ninth House--philosophical and scientific institutions, general knowledge the people have of higher education, religion, commerce in general--especially its ability to expand, cables, long distance communication and travel, legalities

Tenth House--president, king or leader of country, the government itself, national trade and influence, national integrity of a country, national celebrities and those having influence in the country, influence a country has in the world of nations

Eleventh House-Senate and House of Representatives, political climate of the country, legislation, international friendships, goals and objectives of the country, humanitarian causes the country espouses

Twelfth House--hospitals, prisons, crime rate, asylums, reformatories, secret societies, spies and secret enemies, psychics, musical and poetic influence, hidden potential of the country

MUNDANE RULERSHIP

Places of the earth ruled by sign and planets

Air--Aquarius, Uranus-Mercury, Gemini

Brier bushes--Mars, Aries

Clouds--Pisces, Neptune

Coal--Capricorn-Saturn

Erosion--Aries, Mars

Flowers--Libra, Venus

Fog--Pisces-Neptune

Fruit--Jupiter, Sagittarius

Grain--Leo, Sun-Virgo, Mercury

Grass--Virgo, Mercury

Herbs--Mars, Aries (secondary rulership--Virgo, Mercury)

Lakes--Cancer, Moon

Metals--Leo, Sun

Mined jewels--Capricorn, Saturn-Venus, Libra-Sun-Leo)

Mountains--Saturn, Capricorn (Capricorn rules opposites--low and high)

Nut trees--Leo, Sun

Ocean--Scorpio, Pluto

Oil--Scorpio, Pluto

Rivers--Cancer, Moon

Roadways--Gemini, Mercury

Sky--Pisces, Neptune

Soil--Taurus, Venus

Swamps--Pluto, Scorpio

Trees--Jupiter, Sagittarius

Valley--Saturn, Capricorn

Vegetable--Cancer, Moon

Water well--Cancer-Moon, Neptune-Pisces

MUSIC

There are many theories about what musical notes rules what; my theory is simple.

C--Aries

C Sharp--Taurus

D--Gemini

E Flat--Cancer

E--Leo

F--Virgo

F Sharp--Libra

G--Scorpio

G Sharp--Sagittarius

A--Capricorn

B Flat--Aquarius

B--Pisces

Planets:

C--Sun

D--Saturn

E--Mercury

F--Moon

G--Mars

A--Venus

B--Jupiter

C Sharp--Uranus

D Sharp--Neptune

F Sharp--Pluto

G Sharp--Chiron

B Flat--For the unknown planet to be discovered

MUTABLE SIGNS

(Sometimes known as common signs)

Keywords are versatile, changeable, subtle inconstant, sympathetic, humanitarian, methodical, sometimes impractical, comprehending, intuitive, tedious, retiring, mentally flexible and ingenious. This person could be a commentator, counsellor or adviser.

Undeveloped types can be crafty and deceptive and may have ulterior motives.

Gemini is sometimes repetitious, absent minded, expressive, able to communicate with anyone regardless of their station in life and prone to worry.

Virgo is analytical, critical, dependent, good at trouble shooting and attentive to details.

Sagittarius is open and objective in temperament, peaceful, dignified, philosophical, foresighted and understanding.

Pisces is imaginative, poetic, musical,

psychic, spiritual, healing and compassionate, tolerant and given to daydreaming.

Houses:

Third House--relationships (brothers, sisters, neighbors)

Sixth House--health and working conditions

Ninth House--religion, higher education, philosophy, foreign affairs, relatives of marriage partner

Twelfth House--subconscious mind, secret enemies (if any), past life indications

OCCUPATIONS

Occupation is indicated by the Sun sign of the person and the sign on the tenth house of the chart.

Aries--soldier, engineer, fire fighter, police officer, one who does his own thing, pioneering a new area for society, or any occupation that requires original, fearless ideas. This person could also have an occupation where he works with tools or sharp instruments.

Taurus--agricultural worker, musician, artist, actor, designer or tradesman dealing with clothing, confectioner, jeweler, sculptor

Gemini--secretary, writer, school teacher, accountant, traveling salesperson, messenger, lecturer, printer, bookseller and linguist

Cancer--clairvoyants, social worker, sailor, fisherman, dealer in fluids, nurse, homemaker, cook, dietician, child care worker

Leo--athlete, executive, pioneer in entertainment or education, government official, foreman, or any other vocation that requires, power and gambler

Virgo--librarian, statistician, computer programer, scientist, promoter, trouble shooter in any vocation, food therapist, exercise instructor and all that goes with taking care of one's body. Negative occupations are thief, counterfeiter (not necessarily of money).

Libra--diplomat, navigator, actor, artist, judge, schemer who always finds a way, musician, counselor

Scorpio--surgeon, healer, architect, investigator, detective, lawyer, chemist, solicitor, psychologist, soldier, leader in large financial organization

Sagittarius--sportsman, lawyer, philanthropist, teacher in higher education, philosopher, minister, banker, orator and promoter of almost anything

Capricorn--specialist in business, routine worker, organizer, real estate agent, builder, miner and farmer

Aquarius--humanitarian, explorer, aviator, actor, astrologer and like careers, electrician, singer, psychologist and radio or TV technician

Pisces--psychic, any occupation dealing with water, musician, cancer, musician, poet

ORIENTAL ESOTERIC ZODIAC

Aries--Light of Parabrahm . . . the spirit stuff of all creation

Taurus--The Logos . . . collective creator of the universe

Gemini--Spirit Matter . . . the eternal substance of which divine natures are compounded

Cancer--Nascient Mind . . . mind in its original state; the beginning of mind

Leo--Prana . . . the breath, solar force, vitalizing agent

Virgo--Linga Sharira . . . the astral body; the invisible form that patterns the physical body

Libra--Sthula Sharira . . . the dense physical body; the elements which serve as a vehicle for the higher nature

Scorpio--Kama . . . the desire nature; the body which is the base of longings, attachments and appetites

Sagittarius--Kama-Manas . . . the emotional mind; the mind likened to Platonic

Capricorn--Manas . . . the thinker; individual self or incarnating ego

Aquarius--Buddhi-Manas . . . the root of intelligence; the formless divine mind; part of the higher self

Pisces--Buddhi . . . the passive spiritual form; the cosmic intelligence that connects to the individual person

PART OF FORTUNE

The Part of Fortune has been the subject of controversy since it was first introduced into astrology by Ptolemy. However, as with everything else, every discovery in astrology will have pros and cons. Placidus and Negusantius were said to have checked its influence and didn't see any. the Arabians, on the other hand, highly valued its importance in the horoscope. My advice is check it for yourself.

Part of Fortune is said to signify material

influences such as finances, and other material or physical fortune of the bearer. Some astrologers believe it has a physical influence on the health of the native, and some do not. Your author does not believe it has much effect all by itself, but does enhance what it touches, making those things stronger in a positive or negative way.

The Part of Fortune is obtained . . .

Ascendant, plus Moon, minus Sun.

Example: Ascendant 4/29 Scorpio . . .Moon 29/02 Aries and Sun 1/59 Gemini . . .

Take signs through Libra . . . add 4 degree Scorpio . . .

7 04 29+

<u>0 29 04</u>

7 34 33-

change to . . .

7 33 93

2 01 59

5 32 34

translated to . . .

6 02 34

This would make it into the seventh sign by 2 degrees 34 minutes Libra. Hence Part of Fortune would be 2 degrees 34 minutes Libra.

Houses:

First House, Aries--luck through own personality or labors

Second House, Taurus--gain through self employment or earning capacity

Third House, Gemini--prosperity through siblings, community or communications

Fourth House, Cancer--luck with domestic concerns or mother's influence

Fifth House, Leo--success with children, creative talents, and pleasures of love

Sixth House, Virgo--success in employment and health

Seventh House, Libra--success as a partner in either business or marriage

Eight House, Scorpio--prosperity through inheritance, fruits of a marriage or the ability to regenerate oneself and partnerships

Ninth House, Sagittarius--success in legal, religious or philosophical matters and with foreign born people or matters

Tenth House, Capricorn--high position with career, and/or father or authority figure

Eleventh House, Aquarius--success with friends, goals and objectives

Twelfth House,Pisces--psychic abilities, skill in intangible, inspirational expression as in poetry, music, and/or things hidden from outer view

PERSONAL HOUSES

These are the fire houses, ruled by the fiery triplicity. the signs on the cusps of these houses, and planets therein, will influence the nature of the individual as follows.

First House: the body and personality in general

Fifth House: the soul

Ninth House: the spirit and higher mind

PHILOSOPHY OF ASTROLOGY

If you go over the history of nations and the history of astrology you will find that no country, no nation, no race has reached the peak of culture without the civilizing influence of astrology. The study of the stars and planets and the effect of their patterns upon the lives of creatures on earth, human and animal, has intrigued the minds of men and women from the beginning of recorded history. Human beings developed a study and belief in astrology long before the invention of the written word. the earliest written records indicate that astrology was already well established in the minds of men.

Astronomers have called astrology the mad mother of astronomy, and some have called it the mad child of astronomy. However, since modern astronomy is a result of astrological speculation, identifying astrology as the mother of astronomy may be most accurate.

Astrology works because it bears witness to the truths of celestial dynamics. Astrology is a universal truth, and more than a predictive art; it is a philosophy of the universe, expressed physically through a particular technique.

Here is the formula: the spirit of things from the stars; the soul of things from the planets; and the body of all things from the elements. These three divisions, spirit, soul and body-consciousness; intelligence and force-will; wisdom and action are related to the stars (suns), planets and elements.

Man living on the planet Earth receives three kinds of energy into his own nature: spiritual energy from the countless stars, physical energy reflected from the other planets in the solar system, and elementary energy from the physical earth beneath his feet.

PHYSICAL APPEARANCE

Planets . . .

Moon: inclined to heaviness, round face, large and soft eyes, short hands and feet, possible thick neck, usually small bones (Cancer)

Mercury: slender body and face, broad forehead, straight and sometimes long nose, thin lips slender fingers, expressive hands, thin beard, penetrating eyes, broken complexion (Gemini, Virgo)

Venus: short but graceful body, heaviness in middle years, round face, brown or black hair (there can be exceptions), large and expressive eyes, soft and musical voice, vivaciousness (Libra, Taurus)

Sun: strong, well formed but bony body; large and commanding face and forehead;

usually light hair and complexion; men inclined to baldness; clear and direct eyes; clear and beautiful complexion (Leo)

Mars; strong and sometimes stocky body, medium height, often has curly hair, quick and direct eyes, ruddy complexion, lived face when angry (Aries)

Jupiter: large and well formed body, heaviness in older years, strong and well formed chest, high forehead, straight nose, widely-spaced eyes that are kind and understanding despite what the person may be saying (Sagittarius)

Saturn: slender body, angular, large-boned, drooping back with older age, stern features, small eyes (Capricorn)

Uranus: slender body, pleasing appearance, large or expressive eyes, brilliant looking, keen, light complexion, aware of things others are not, erratic in movements (Aquarius)

Neptune: long head, retreating hair at temples, hypnotic yet gentle and tolerant eyes, appears to be listening to something others are not--such as music, finely put together body (Pisces)

Pluto: dark and penetrating eyes, sexually attractive, strong body, strong voice, direct walk--as if their walk is making a statement (Scorpio)

Signs . . .

Aries: long head and face, angular check bones, high temples, broad chin, narrow and thin mouth, body slender

Note: These characteristics may be true of the Sun sign but are especially true of the ascendant which rules the body.

Taurus: short head and neck; broad and somewhat flat face; strong chin; large eyes; usually dark hair; full lips; short, sturdy and plump body

Gemini: long head and neck, broad forehead, pointed chin, expressive eyes, light hair, thin lips, small mouth, slender body, long fingers

PHYSICAL RULERSHIP

Aries--head, face, eyes, cerebrum, upper jaw

Taurus--neck, cerebellum, upper cervical vertebra, ears, throat, larynx, eustachian tubes, tonsils, upper part of esophagus, palate, thyroid gland, vocal cords, lower jaw

Gemini--arms, shoulders, muscles and bones of arms and shoulders, lungs-including trachea and bronchi, thymus gland, upper ribs, capillaries, hands

Cancer--stomach, armpits, breasts, lacteals, chest cavity, solar plexus, pancreas, epigastric region, diaphragm and upper lobes of liver, and to some extent the womb

Leo--heart, spinal column-with its marrow-especially dorsal region and spinal cord

Virgo--intestines, alimentary canal, abdominal cavity and its membranes, spleen, lower lobs of liver

Libra--kidneys, ovaries, adrenals, loins, appendix, ureters, lumbar vertebrae and contiguous areas, skin in general

Scorpio--generative organs, nasal bone, gall bladder, pubic bone, lower lumbar vertebrae, prostate gland, testicles, colon, rectum

Sagittarius--hips, thighs coccygeal vertebrae, sacrál region, sciatic nerves, blood

Capricorn--knees, kneecaps, hair, outer epidermis, various joints of the body, bones in general

Aquarius--lower leg--including the calves and ankles, teeth, blood circulation in general

Pisces--feet and toes-especially their bones and muscles, generative organs, the lungs (by sympathy)

PHYSIOLOGY

Earth/Taurus--assimilation and excretion of food

Moon/Cancer--tears, fluids, impregnation, dissolution, body rhythm, flow of all secretions, relaxation

Mercury/Gemini--nerves, respiration, sense of touch, interaction of parts, motion

Venus/Libra--flow of bodily fluids, relaxation of muscles, filtration (kidneys)

Mars/Aries--energizing, swelling, heating, control of metallic elements in the body, eruptions

Sun/Leo--generative force, physical growth, circulation, expansion

Jupiter/Sagittarius--molecular nutrition, cell development, blood building, flesh, parts working together

Saturn/Capricorn--congestion, cartilages, bones, accumulation of bone deposits, ossification

Uranus/Aquarius--magnetism, function during sleep, psychological action, electrical forces, activity going on in the body

Neptune/Pisces--kundalini function, motion of chakras, spiritual reaction on body, dissolving of any muscular or bone deposits

Pluto/Scorpio--regeneration, generative functions, sex, tumors, dissolving of tumors, general healing of the body

Chiron/Virgo--general redirection of health, the ability to discipline body to make a perfect whole

PLANETS AND SIGNS

You do not need a cookbook definition of the planets in the signs. The Sun is a person's ego (a main influence). When the Sun is in Aries, for instance, the ego will be defined like Aries. If the Sun is in another sign, the ego will be defined like that sign.

The Moon is the emotions and feeling, the mother, the nurturing instincts, and the woman in a man's chart. many astrologers think a woman is more like her Moon sign than her Sun sign. If a person's Moon sign is in Leo, for instance, the emotions and feeling nature will be like that sign.

Mercury stands for subjective conscious thinking, and for one's communication on that subjective level. When Mercury is in the sign Virgo, for example, the thinking will be geared toward Virgo influence.

Venus stands for love, what the person values, and his method of showing love. Whatever sign Venus is located in is how he is going to do that.

Mars stands for sexual expression and how one gets what one wants. Whatever sign mars is in is how that energy is utilized.

Jupiter is the planetary influence of objectivity and how expansion and opportunity is expressed. Whatever sign Jupiter is in is how that happens.

Saturn is the planetary influence of discipline and responsibility and authority. Whatever sign it is located in is the expression of this responsibility.

Uranus is the planetary influence of individuality, creativity, and where and how one expresses it. the sign is the vehicle by which that happens.

Neptune is the influence of the intangibles, compassion, music, poetry, and beliefs. the sign that Neptune occupies is the Neptune belief.

Pluto is the planetary influence of regeneration, rejuvenation, healing, death, and rebirth. The sign Pluto is located in is how and why that happens.

Chiron is the influence of karmic breakthrough. The sign it occupies is where and by what this happens.

North Node represents new growth in this life. The sign it occupies defines that new growth.

South Node is growth and learned talent from another life. The sign represents that talent.

Earth is what is in the shadow of the Sun. It is the person's foundation and strength. The sign is the location of that foundation and strength.

Ascendant is the time of day the person was born. The sign of the ascendant is the physical body of the person, and the personality he projects to others.

Midheaven is the highest point in the chart; its sign represents occupation and public image.

Juno, an asteroid, is cosmic marriage; and what the person needs to fulfill in body, mind, and spirit in a relation ship. The sign of Juno shows that need and influence.

Pallas, and asteroid, is cosmic wisdom. Its sign shows what that wisdom is.

Vesta, an asteroid, is where there is a small, constant, brilliant light ready to respond and grow when necessary. The sign is how and why that might take place.

Ceres, an asteroid, is where one can give

unconditional love and devotion to something or someone, and where, at some point, the person receives a reward. The sign of Ceres is how, why and what that is.

Part of Fortune by sign is where luck can be obtained.

The house position of these planets and signs is where experiences can come to that planet to fulfill a day to day existence in this life. For example: Sun in Gemini in the seventh house (marriage and deep relationships) is the experiences the Sun in Gemini has to complete in its mission in this life.

Moon in Aries in sixth house is a emotional nature that leads in the area of health knowledge; and it also inspires people to "do their own thing: in seeking fulfillment in emotions and working conditions on a daily basis.

PLANETARY CONDITIONS AFFECTING PEOPLE

Saturn produces positive discipline or

scarcity.

Jupiter produces abundance or extra-vagance.

Mars produces productive activity or arguments or contention.

Venus produces peace, monetary well being, or lack of money, or peace.

Mercury produces positive communication, educational advantages and good or bad changes, depending upon the aspects.

Uranus produces inventive ideas, humanitarian benefits, erratic actions, and undependable situation.

Pluto produces healing techniques, regenerative opportunities, discovery of hidden facets; negative influences produce domination, deaths due to disease, and violent crimes.

An ancient astrologer Pythagoras predicted there were ten dots representing the energies of the solar system. At the time he said this, the astrologers only recognized seven planets. Three have been added since that time, producing ten (including the Sun).

Peculiar happenings are often noticed for which there is no legitimate astrological explanation. Perhaps there are powerful energies yet undiscovered.

While there will always be doubts concerning the invisible elements of our solar system, it is the responsibility of the astrologer to deal with just the known planets at the time of the construction of the chart.

Pluto is a relatively recent astronomical discovery. Most astrologers agree it has higher octave of Mars. Some believe it is the planetary influence that holds the "Secret Doctrines" and those things that have been mysterious to us in the past. In time we will find the answer to mysteries we have long sought.

PLANETARY REVOLUTIONS

(equivalent to our year-the path around

the sun)

Earth--365 1/4 days

Moon--See PLANETARY ROTATIONS

Mercury--88 days

Venus--225 days

Mars--1 year, 322 days

Jupiter--11 years, 317 days

Saturn--29 years, 168 days

These are the inner planets that deal with the seven phases of life.

Uranus, Neptune, Pluto have a revolutionary period beyond the average lifetime.

PLANETARY ROTATIONS

(equivalent to our day)

Earth--24 hours

Moon--27 1/4 days to same position (with respect to stars), 29 1/2 days to same phase (with respect to Sun)

Mercury--58 1/2 days

Venus--243 days

Mars--24 1/2 hours

Jupiter--10 hours

Saturn 10 1/4 hours

Sun--25 days (at equator)

PLANETARY RULERSHIPS

Moon--daytime-Cancer, nighttime-Cancer

Mercury--daytime-Gemini, nighttime-Virgo

Venus--daytime-Taurus, nighttime-Libra

Sun--daytime-Leo, nighttime-Leo

Mars--daytime-Aries, nighttime-Aries

Jupiter--daytime-Sagittarius, nighttime-first 15 degrees of Pisces

Saturn--daytime-Capricorn, nighttime-first 15 degrees of Aquarius

Uranus--day and nighttime-Aquarius

Neptune--day and nighttime-Pisces

Pluto--day and nighttime-Scorpio

PERSONAL HOUSES

These are the fire houses, ruled by the fiery triplicity. The signs on the cusps of these houses, and planets therein, will influence the nature of the individual as follows.

First House--the body and personality in general

Fifth House--the soul

Ninth House--the spirit and higher mind

PRE-NATAL EPOCH

We all know that counting nine months backward from the actual physical birth, we can calculate the approximate time of conception. One theory holds that the Ascendant at birth time was the Moon sign on the day of conception.

I figured my conception must have taken place around the first few days of September (my birthday is May 23). I believe I was a couple weeks early, according to the calculations my mother made.

It seems very interesting that most of the major things that have happened to me in my life happened the first two weeks of September and not my solar return. Every major event in my life seemed to take place in September (Virgo time). Calculate back to your time of conception and test for yourself what has happened of importance in your life at that time.

Outside of that, do your own investigating. Chances are, any major event in your life has occurred close to the same time of year.

PROGRESSIONS

The sun sign: one day equals one year, two hours equals one month, thirty minutes equal one week, four minutes equal one day, one minute equals six hours, ten seconds equals one hour and one second equals six minutes.

The first day will show the person's first

year. The second day after birth will show by progression the second year, and so on. The twenty-fifth year will show by twenty-five days after birth. Whatever aspect the Sun makes to natal planets is a more or less major point in time.

The Moon progresses a degree each month. It is a faster progressed influence. Every degree of the Moon after birth represents a month. Whatever natal planet this progressed Moon aspects shows is a possible influence of that aspect. The Moon is the fasted progressing influence.

The Ascendant and Midheaven progress at approximately one degree per year. Since the Ascendant is personality and body, when a progressed aspect comes about there could be a change in the physical body, appearance, or personality according to the nature of the progressed aspect.

The same is true of the progressed Midheaven, where a progressed aspect could bring a new influence to the career or public image.

Astrologers disagree about whether progressed planets aspect other progressed planets. Thus, researchers will have to make up their own minds about this particular issue.

One thing all astrologers agree on: when progressed planets aspect natal planets in the chart they trigger certain influences in the person's life. They makes the person aware of his or her talents, potential and timing.

Four hours, for example, after the birth of child, whatever aspects planets make at that time will be felt two months after birth.

PROPHECY

In looking to the future, one has to consider more than one transit or progression influencing the chart. Man has his own free will and is not simply being controlled by the destiny of the stars. Part of the influence is how he was brought up as a child, which helps determine which way the influence is going to affect him.

Nothing in the future is crystallized. They only impel a person toward a potential of learning and development. the analogy I use is, "If you know there is a hole outside your door you will not fall in it. I you don't know it is there, you just may."

Probably the strongest influence on us is the Moon, since we are often moved by it's emotional impact.

The Sun progresses approximately a degree a year. The Moon progresses approximately a degree a month. when these progressions touch a natal planet by degree, it touches off a new phase for the person.

The transits are the current positions of the planets, Moon and Sun. How are these current positions influencing the natal chart of the person? Each planet has a unique influence according to its nature, as we will see later.

QUICK ANALYSIS OF THE CHART

1) Rising sign: a one word definition of the personality and body

2) The Sun: a one word definition of the ego

3) Midheaven sign: a one word definition of the career

4) Sign of Moon and house position: a two word definition

5) All the other planets: check sign and house positions and give one word definitions of those positions according to the nature of the planets.

READING A CHART

1) What area of life are you looking to interpret?

2) What is the sign on the cusp of the house?

3) What is the decanate of the sign?

4) What house and aspects does the ruler of the sign have?

5) What aspects do the ruler of the sign and the ruler of the decanate make?

6) What house does the ruler of the decanate occupy? This is an outlet for the area of life you began checking.

7) Are there any other planets in the house in which the ruler of the decanate is located?

8) Are there any other planets in the house in which the ruler of the sign is located?

9) Check every aspect for all potential influences in the area of concern.

RELATIONSHIP HOUSES

The relationship houses are ruled by the air signs. The signs on the cusp and the planets therein describe attitudes with relationships.

Third House--blood relations and neighbors

Seventh House--marital relations and business partners

Eleventh House--friendships, social and mental relationships

SEVEN STAGES OF MAN

When man is born into this world he goes through seven phases.

Moon--(and its sign and house) consists of the first four years when there is emotional nurturing. this is "growth."

Mercury--(and its sign and house) consists of ten years from the fifth year through the fourteenth year when man is concerned with his mind and his environment. The word is "education" (formal and self taught).

Venus--(and its sign and house) consists of eight year, from the fifteenth year through the twenty-second year. This is the development of love and caring for others. The key word is "love." The person aspires to connect with love.

Sun--(and its sign and house) consists of nineteen years, from the twenty-third year through the forty-first year. He finds his place in the world and is vigorous and wants to shine. The key words are "ego placement in the world."

Mars--(and its sign and house) is the fifth phase and consists of fifteen years. This is from the forty-second year to the fifty-sixth year. This person is sexually vigorous and very active in his world. He aspires to and is proud of his manhood, wherever that takes him.

Jupiter--(and its sign and house) is the sixth phase and consists of twelve years, from the fifty-seventh year through the sixty-eighth. The person has gained knowledge and is prudent in projecting authority (or should be). The key words are "reflection of what he has accomplished." By this time he will have advanced his position in the world according to his talent.

Saturn--(and its sign and house) is the seventh phase and is from age sixty-eight to the end of one's life. The person begins to resign himself to the rest of his life and begins to see a physical decline (according to how he has or has not taken care of himself). By this time he is either considered wise or cranky.

Uranus, Neptune and Pluto are not included in the seven phases since they are not concerned with that part of man that experiences change or decay, but relate to the spiritual nature, creative intellect and hidden potential. Most people respond to the first seven phases. When a person is very intellectual, spiritually responsive to the universe, he incorporates these three planetary influences in his daily life almost from the beginning.

SOLAR ECLIPSE

Solar eclipse is when the Moon is between the Sun and the Earth and the Moon casts a shadow on the Earth. This can be partial and total.

The belief that eclipses are always destructive in their influence should not be taken seriously. An eclipse does represent tremendous power, but if it is well aspected it can signify great good. Most astrologers only describe the bad influence, which is unfair and very inaccurate.

When an eclipse forms a trine or a sextile the emphasis will be fortunate and powerful. Squares and oppositions can bring misfortune according to the planet, ascendant, or Sun it touches. Eclipses do not take effect immediately; sometimes several months pass before a situation arises which the eclipse has put into motion. Therefore, eclipses are sometimes misinterpreted. For this eclipses are hard to define and interpret. Yet they do influence us one way or another.

If an eclipse is in a fixed sign the effects last a very long time. If it is in a cardinal sign, they will last a short while but be active. If it is in a mutable sign, the effect will be sporadic. When an eclipse is in a fire sign, it can be accompanied by fire, destruction and war. On the positive side, it may activate something long needed. When it is in an earth sign, droughts, famines and earthquakes may occur but will be accompanied by positive, practical solutions. When in a water sign, the person may experience sensitive, emotional outbreaks, but also sensitive awareness. When an eclipse happens in an air sign there may be disruption of communication--either nationally or personally. The positive side of an air eclipse can be greater understanding between people, and mental awareness.

SPIRITUAL TEMPERAMENT

Moon/Cancer--nourishment, magnetism,

a sense of mother-God's existence, fecundity, physical beauty of animal man

Mercury/Gemini-Virgo--communication, spiritual human consciousness, the reason for life, first level phases of Virgo intuition and magnetism, Gemini electricity

Venus/Libra-Taurus--love, values, unification with another, beauty of love and caring, value of our material world; Libra electricity, Taurus magnetism

Sun/Leo--universal consciousness, life giving, instincts of father-God, electricity, creativity

Mars/Aries--consciousness of energy and direction , the spiritual pioneer who dares, electricity

Jupiter/Sagittarius--higher levels of intuition, he who sees higher and wider possibilities, idealization, the uplifter, appreciation of vegetative poser, nobility of purpose that encompasses more than the material, electricity

Saturn/Capricorn--crystallization,magnetism,endurance, understanding of self, subdued nature, obstruction for a better purpose, the tester, the authority

Uranus/Aquarius--perfect spiritual consciousness and understanding, the enlightener, altruististm, both electricity and magnetism (androgynous)

Neptune/Pisces--the dissolver, consciousness of the intangible, or of the wider spiritual divinity, magnetism, sacrificial payment

Chiron/Virgo--correction of karma, escape into a wider and more productive elevation

SYMBOLISM

Cardinal, fixed, and mutable represent the triplicities that consist of the four elements of nature--earth, water, fire and air. The elements were thought, when combined, to produce form. the triplicities signify the threefold constitution of both man, and the universe in which he lives. When you combine the triplicities and the four elements you have the seven-fold symbols, or parts, of animals, plants, and other living entities.

Earth represents the physical nature of all things. Water represents the emotional nature of all things. Air represents the mental and supermental nature of all things. Fire represents the spiritual nature of all things and the unfolding of latent talents and potentials.

The three signs that form each triplicity are in trine to each other, in other words, 120 degrees apart. Each sign is 30 degrees. It has been an astrological teaching since ancient times that people should marry someone in their own triplicity in order to be in harmony with each other: air to an air sign, water to a water sign, fire to a fire sign and earth to an earth sign.

SYMBOLS

(electric-magnetic)

The ancients used certain symbol to show

the personal nature of a particular sign.

HUMAN signs were meant to show intellectual and communicative talents above and beyond mundane physical needs. Sometimes Libra was included in the list.

ANIMAL signs are the signs of the zodiac that have animals as symbols. These signs are Aries, Taurus, Leo, Capricorn, and the last half of Sagittarius. This increases their passion for physical excess and/or material things, like money, food, and other things that satisfy the physical body.

REPTILE signs are Cancer, Scorpio and Pisces. They are secretive, sometimes tricky, and not easily understood by others. They do not completely expose their talents at any given moment, but instead wait until the time is right to bring them out in the open.

Signs that have a TWO-FIGURE symbol are Gemini, Sagittarius, and Pisces. This means they have dual natures almost like two souls in one body.

The ancients did some rather nasty interpretation by modern standards. They called every other sign starting with Aries masculine signs; these were supposed to be fortunate because they were male-oriented. Every other sign starting with Taurus was supposed to be misfortunate. It was interpreted this way because in ancient times, and even in the early Christian era it was regarded a misfortune to be for a woman. Stupid and silly, yes, but that is one of the fallacies of early astrological interpretation.

Although Aries, Gemini, Leo, Libra, Sagittarius, and Aquarius are thought of as masculine signs, they are really just electric and outgoing.

If one eliminates the early, ancient interpretations from astrology, which were biased, you will be left with the truth.

TERMINAL HOUSES

These are ruled by the water signs. The

signs on the cusps, and the planets therein, indicate terminations and/or regenerative influences of life.

Fourth House--the end of the bodily man (how, when and why)

Eighth House--the liberation and condition of the soul

Twelfth House--the philosophical or second death, or awakening

TRAITS

There are negative traits of the astrological signs/planets as well as positive. It is important that any student of astrology knows what they are. One cannot say Gemini is such and so in positive definition and Leo is such and so in a negative fashion. All astrological signs have both.

CANCER/MOON:

positive--magnetic, psychic, peaceful, ingenious, nurturing all of life

negative--frivolous, procrastinating, lazy

selfish, materialistic, clinging to those loved, jealous

MERCURY/GEMINI-VIRGO:

positive--humorous, dexterous, brilliant, sensitive, pleasing in conversation, literary, aware of surroundings

negative--gossipy, prone to gambling, profane, forgetful, worrying, nervous, unprincipled, sometimes conceited

VENUS/TAURUS-LIBRA:

positive--cheerful, amicable, affectionate, friendly, artistic, suave, harmonious in most things, able to see both sides of a situation

negative--sentimental, vain, superficial in expression, lustful, lazy, prone to excess in the good things of life

SUN/LEO:

positive--dignified, vital, ambitious, dramatic, loves children and education, creative negative--pompous, loves display, egotistic, lacking in ambition because of excess in the pursuit of pleasure

MARS/ARIES:

positive--brave, impulsive, gallant, enterprising, expert in taking action on original ideas, pioneering, fearless, independent, energetic.

negative--destructive, overly passionate, coarse, cruel, warlike, unable to relate to others, sarcastic with those who do not think like he does

JUPITER/SAGITTARIUS:

positive--benevolent, honorable, humorous, understanding, patient, wise, just, popular, philosophical

negative--extravagant, procrastinating conceited, self-indulgent, hypocritical, irresolute, lazy

SATURN/CAPRICORN:

positive--faithful, systematic, tactful, thrifty, cautious, responsible, disciplined, punctual, studious, sensible, chaste

negative--too crystallized, pessimistic, suspicious, jealous, timid, stern, fatalistic, too traditional

URANUS/AQUARIUS:

positive--compassionate, intuitive, creative intellectual, scientific, enthusiastic, prophetic, heroic for friends or the people in general, unconventional in a positive way

negative--eccentric, bohemian, irresponsible, unpredictable, self-centered, stubborn, unconventional in a negative way, hard to put into a controlled system

NEPTUNE/PISCES:

positive--spiritual, penetrating, psychic, romantic, poetic, musical, inspiring, prophetic, fascinating

negative--morbid, fraudulent, unreliable, excessively erotic, deceptive, self-pitying, alcoholic, prone to sensual excess

PLUTO/SCORPIO:

positive--good at research, heroic, healing, has a deep understanding of the nature of life, able to have soul love

negative--dominating, secretive, destructive, selfish, coveting others luck (jealousy), vindictive

CHIRON/VIRGO:

positive--able to find a new pathway, an outlet, a breakthrough in karma, where one can change a bad situation

negative--fault-finding with others instead of self, dwelling on the tiny speck and magnifying that speck into a mountain

TRANSITS

Transits are the current positions of the Sun, Moon and planets. One examines how the current positions of the planets influence one's natal chart. See what house they fall in and what aspect they make.

The transits of the Moon have a daily influence. Transits of Mercury and Venus also have a shorter, more minor influence. Transits of Jupiter and Saturn last about a year and have a more powerful influence on the natal chart. Transits of Uranus, Neptune, and Pluto move slowly, so they effect a given natal planet for longer time. This gives them a major influence.

The difference is progressions are the person's own timing. Transits are current condition, and come more from outside sources, such as people, events, or situations the person has to deal with in the time span considered.

Mercury--communication, siblings, environment, community

Venus--love, money, caring

Jupiter--expansion, understanding, breaking out of a bind

Saturn--responsibility, discipline, authority figures, father

Uranus--change moving, creativity, individuality, unconventionality

Neptune--psychic posers, music, faith, intangibles, sacrifice

Pluto--transformation, metamorphosis, healing, sometimes death

WATER SIGNS

Water signs are the emotional signs.

Cancer represents nurturing and is concerned with the latter part of one's life and with domestic needs.

Scorpio is concerned with death and/or rebirth.

Pisces is the spiritual part of our nature. It controls rebirth and the karma of our spirit and soul through our reincarnations.

Water signs are usually victims of circumstances.

They can be altruistic and profound. They are the signs who are the most concerned with natural law of the universe.

Because water has a fluid state, people in water signs are easily influenced, and susceptible to moods and psychic forces.

Water has three states: solid (ice), gas (vapor), and liquid. Cancer represents water in its fluid state and is thus subject to changes from without. Scorpio is frozen and the least responsive to moods. It is mysterious and secretive. Pisces is water in the vaporous state like mist or steam. This sign is the bridge between the physical and the metaphysical.

WHAT IS AN AGE

We keep hearing about the New Age. This results from the precession of the equinox. The gyratory motion of the Earth's axis of rotation describes a circle with a radius of 23 1/2 degrees to the plane of the ecliptic. It takes 26,000 years for it to make an entire circle, or entire precession starting at the first point of Aries. This precession amounts to 55 seconds a year. That means it takes 2150 years for an age. The age we are now going into is Aquarius. We are coming out of the age of Pisces. The twelve constellations of the zodiac are found along the ecliptic which is the Sun's apparent path among the stars. This collection of constellations has always been known as the zodiac because of the animal outlines they seem to make. Taken in their natural order they are Aries, the Ram; Taurus, the Bull; Gemini, the Twins; Cancer, the Crab; Leo, the Lion; Virgo, the Virgin; Libra, the Scales; Scorpio, the Scorpion; Sagittarius, the Archer; Capricorn, the Goat; Aquarius, the Water Bearer; and Pisces, the Fish.

Even though we continue to say that the Sun enters the First point in Aries at the beginning of spring, in fact, because of the recession, it has been in Pisces and will soon be in Aquarius. Astrologers and astronomers disagree about the rightful influence (if any) of these zodiacal constellations, and their current positions by precession.

Perhaps one day we will be able to put it

all together and favor investigation rather than limitation. Also maybe we should start investigating EARTH centered astrology. Spring conditions describe Aries. Taurus is described in sequence after Aries, and January describes the personality of Capricorn, etc.

ZODIAC AGES

The precession of the equinoxes establishes the ages. Each age is approximately 2,160 years.

The age just closing is the Piscean age. It has been the age of sorrow, persecution, illusion, drugs, alcohol, and the oppression of the weak by the strong in order to push the erroneous beliefs of the Piscean Age. Many of the occult secrets were hidden in the temples of antiquity to safeguard them from Christian persecution.

The task during the Aquarian Age will be spiritualizing civilization according to the total cosmic pattern, and bringing things back together that were separated: astrology, science, spirituality and our true physical destiny. We were meant to achieve super-physicality but the church has negated it as sinful and bad and stopped it.

The Aquarian Age is also the age of man, of equality among all men and women. Jesus, himself said, "At the end of the Age (Pisces) the truth will be revealed." He too was misinterpreted and misread.

ZODIACAL DESIGNATIONS

Capricorn: December 22 - January 20

Aquarius: January 21 - February 19

Pisces: February 20 - March 20

Aries: March 21 - April 20

Taurus: April 21 - May 20

Gemini: May 21 - June 20

Cancer: June 21 - July 22

Leo: July 23 - August 22

Virgo: August 23 - September 22

Libra: September 23 - October 22

Scorpio: October 23 - November 22

Sagittarius: November 23 - December 21

ZODIACAL DESIGNATIONS WITH OPHIUCHUS

Capricorn: January 19 - February 17

Aquarius: February 17 - March 13

Pisces: March 13 - April 19

Aries: April 19 - May 15

Taurus: May (14)15 - June 21

Gemini: June 21 - July 21

Cancer: July 7 - August 11 (some of these constellations overlap)

Leo: August 12 - September 17

Virgo: September 21 - November 1

Libra: November 1 - November 23

Scorpio: November 23 - November 30

Ohpiuchus: November 30 - December 18

Sagittarius: December 19 - January 19

Ophiuchus is a vast constellation which is in fact a zodiacal constellation. It is north of the zodiacal constellation Scorpio.

Ophiuchus is supposed to be the doctor, the healer. It is also known as the serpent holder, that influence which can control any negativity or the taming of violence. It is channeling an energy into productivity. At least that is the meaning at its best.

So figure out where you are and see if you fit the new sign that has resulted from the precession of the equinox.

ASTRONOMY

ACTUAL SUN POSITION IN THE SIGNS - See Part 1

ANNULAR ECLIPSE

When the Moon is at apogee (farthest away) its apparent diameter is less than the Sun so that the Sun does not completely blocked, but has a bright ring around it.

ASTEROID

Sometimes called minor planets the name asteroid was suggested by Herschel because they resemble stars in appearance. See also Part 1.

ASTRONOMICAL AWARENESS

What is a planet? It is a body or mass which revolves around the Sun and is held in orbit around that Sun. Earth is one of those planets. Several planets have been discovered by some disturbing influence on the position of a known planet.

Planets can be seen in the night sky

because they reflect light from the Sun. Five are visible to the naked eye-- Mercury, Venus, Mars, Jupiter and Saturn. The first time I saw Saturn through a telescope was a memorable experience, for the rings around the planet Saturn are majestic.

These five planets were known since ancient time. Uranus was discovered by accident in 1781. Neptune was discovered in 1846, and Pluto in 1930 However, these are only the larger planets. Since 1801 about 3000 smaller bodies have been discovered revolving around our Sun; they are the small minor planets including Chiron and the asteroids. These small planets lie between Mars and Jupiter.

Venus and Jupiter are often confused in sightings because they look similar to the naked eye. Venus, however, is brighter than Jupiter. During six months of the year it is a morning star and for six months, an evening one.

Mars is the red planet, and is easily

recognized by its color.

The orbits of the planets in order of increasing distance from the Sun are Mercury, Venus, Earth, Mars, Jupiter, Saturn, Uranus, Neptune, and Pluto. (Due to its eccentric orbit, Pluto sometimes slips in front of Neptune.)

ATMOSPHERE OF THE PLANETS

Mercury--tenuous atmosphere

Venus--enveloped by a thick layer of clouds

Mars--a slight atmosphere consisting of carbon dioxide with some water vapor

Saturn--nitrogen and methane, clouds which show ammonia

Jupiter--clouds with ammonia

Uranus--dense atmosphere, mostly methane

Neptune--dense atmosphere, mostly methane

Pluto--too little is known

BINARY

Two stars which revolve around a common center of gravity.

BRAHE, *Tycho* (1546-1601)

Brahe was a Danish astronomer. Before the invention of telescopes apparent star positions could be found only by sighting with an alidade, and aligning the star with two fixed points, such as a wire and a small hole. Brahe was an expert not only in his observations, but also in making his own instruments. He had a famous observatory called Uraniborg and Stellborg, built on the island of Hven near Copenhagen. This was a gift of Frederick II, King of Denmark. Because of Brahe's teaching, his pupil Kepler was able to formulate the laws of planetary motion. Brahe and Kepler worked together for a very short time in Prague before the former's death. Brahe's birthday was December 23, 1546. Kepler's birthday was December 27, 1571.

CALENDAR

The year corresponds to a recurrence of the seasons. It contains nearly 400 days, but some people felt a smaller intermediate unit was needed--the month. the numbers most easily divisible by several factors was the number twelve. I early days the month was described by the lunation of the Moon. In our calendar year there are twelve months and thirteen lunations. A week of seven days was chosen to divide the month. The number seven has long had special significance. The Greeks after various trials and errors adopted a calendar of 365 days distributed over twelve unequal months. The beginning of the year was the spring equinox, which seems more logical than the January 1st date we observe now.

The months of July and August were so named to honor two great emperors. Julius Caesar, with the assistance of the astronomer Sosigenes, devised the calendar still in common use. Three years out of four had 365 days, the fourth year had 366 days (hence leap year) . . . February 29.

CALLISTO

Fifth satellite from the planet Jupiter. It was discovered by Galileo in January 1610.

CANALS - Mars

They are a network of lines on the surface of Mars. The existence of these channels was announced by the Italian astronomer Schiaparelli (1835-1910) in 1877. However, over the years there has been much dispute as to the interpretation of what they really are. Erroneous translation from the Italian labelled them as "canals."

CAPRICORN

Capricorn, the goat, is the tenth constellation of the zodiac. Its most important stars do not exceed third magnitude.

CASSINI, *Giovanni Domenico* (1625-1712)

He was a distinguished astronomer. He was descended from an Italian family, studied in Genoa, and became a university professor at Bologna. He was one of the most outstanding observers in a period when little attention was paid to the physical nature of the heavenly bodies. He discovered four of Saturn's moons: Tethys, Deone, Rhea, Japetus.

CELESTIAL BODY

A body external to the Earth, which can be seen by the naked eye.

CELESTIAL MECHANICS

Motion under the influence of force. The conditions under which motion takes place.

CELESTIAL POLE

The points on the celestial sphere directly above the terrestrial poles.

CENTER OF GRAVITY

It is that point in a solid body at which gravity has the strongest power.

COMA BERENICES

It is a large constellation in the northern hemisphere, situated north of Virgo, and between Bootes and Leo. This constellation contains the north galactic pole. Since there is very little light absorption by interstellar matter in this region, an exceptional number of galaxies can be seen. This system of stars extends into the neighboring constellation of Virgo.

COMET

Newton was the first astronomer, with the help of Halley, to become interested in comets. He was able to show that those seen in 1531, 1607, and 1682 were one and the same object. It now bears Halley's name.

The orbits of comets are elliptical, but come in various sizes . They have one characteristic in common: great eccentricity. Comets can be seen for only a short period of time, usually when near the Sun.

Comets consist of tenuous gas emanating from a solid nucleus.

They belong to the solar system. There is no evidence of a comet coming into the solar system from the outside. When it first appears, it is small and round; as it comes nearer it becomes elongated by developing a tail. The tail may be of great length, and it provides a great sight for the naked eye. The tails are not necessarily straight; some are curved. Sometimes people call comets shooting stars but this is not the case. Shooting stars are meteors. They are most frequently seen when swarms of meteors make an annual appearance. Meteors described later in this astronomy section.

CONSTELLATION

For as long as man has existed on Earth,

the positions of the stars have changed only very slowly. Ancient patterns in the sky were thus similar but not identical to modern ones.

We have the pattern of stars known as the Great Bear, the Northern Crown, the Lyre, Orion, the zodiacal constellations, and many more. There are 88 constellations as recognized by the International Astronomical Union.

COPERNICUS, *Nikolaus* (1473-1543)

He was a Polish Astronomer. He was the son of a rich merchant. He studied astronomy at the University of Krakow (1491-1495), then at the University of Bologna (1496). In 1501 he was elected a canon of Frauenburg Cathedral. Having obtained permission to continue his studies in Italy, He enrolled in the faculties of medicine and law at the University of Padua, but at the same time continued his astronomical observations. Throughout his life astronomy was his favorite study, and his name has become immortal as a result. Contrary to the ideas then current, he proved that the Earth does not occupy the central position in the universe, but, along with the other planets orbited the Sun. His system was fully worked out in 1515, but it was not until 1543 that he decided to publish his works at Nurnmberg--his immortal treatise on his theories for the world to read and to know. His birthday was February 19, 1473. His Sun and Mercury were in Pisces; Moon and Jupiter were in Sagittarius. Mars was in Aquarius and Saturn was in Pisces. Venus was in Aries.

CORONA BOREALIS

This is a constellation of the northern hemisphere situated between Boötes and Hercules; easily recognized by its incomplete circle shape.

CORONA, SOLAR

It is a gaseous halo around the Sun, seen

for a brief period during a total eclipse.

COSMOLOGY

This represents the study of the universe as a whole, its origin, and evolution.

COSMOS

This is a word used to represent the universe as a totality.

CRAB NEBULA

It is a small planetary nebula in the constellation Taurus. The Crab Nebula is the remains of a star that exploded as a Supernova in 1054. It was the first object in Messier's catalogue, designated as M-1.

CURVATURE OF SPACE

Space, in the normal description, seems to have three dimensions. Einstein introduced the concept of space-time in four dimensions, in which a gravitational field shows as a curve in space-time.

CYGNUS

One of the brightest constellations visible in the summer, Cygnus the swan, which lies close to Lyra, is also known as the Northern Cross.

DAY

This is the time taken by a planet to rotate on its axis.

Sidereal day . . . the time taken by the earth to rotate on its axis with **respect to given stars**. It defines the appearance of stars in the sky at any moment anywhere on the planet . . . and the time taken to return to the meridian of observation.

Solar day . . . the time for the Sun to return to the meridian at the place of observation. True solar day relates to the return of the true Sun while the mean solar day relates to an imaginary Sun whose motion is uniform. Human affairs on Earth are regulated by the mean solar day; one result of this is that the twelfth hour falls in the middle of the day.

DECLINATION

The angle between the direction of the star and the plane of the celestial equator is the declination; this is determined 0-90 degrees, positive for stars north of the equator and negative for those south. In other words north declination and south declination.

DEGREE OF LONGITUDE AND LATITUDE

The Earth is divided into 360 degrees of longitude (east-west) and 180 degrees of latitude (north-south). Lines of longitude get closer together near the poles; latitude lines are equidistant from pole to pole.

DENEB

The star at the tail end of the constellation Cygnus, the swan also called The Northern Cross.

DIONE

The fourth satellite or moon of Saturn.

This was discovered by Cassini March 21, 1684. It is one of Saturn's satellites that can be seen without much difficulty.

DWARF STAR

Medium-sized star, also known as a main-sequence star, that is intermediate in size between giant stars and white dwarfs. Size varies from one tenth Sun's diameter to ten times.

Ninety percent of stars as seen from Earth are dwarf stars. The Sun is a dwarf star.

EARTH

From the point of view of astronomy, the Earth is one of the planets in the solar system.

Earth's atmosphere has the following composition . . .

nitrogen (78 percent)

oxygen (21 percent)

argon (less than one percent)

carbon dioxide (less than one percent)

hydrogen (one hundreth of a percent)

neon (one thousandth of a percent)

helium (less than a thousandth)

krypton (trace)

ECLIPSE

Total or partial disappearance of a celestial body in the shadow of another. During solar and lunar eclipses, as seen from the Earth, both the Sun and the Moon have approximately the same apparent diameter of one half degree. In a solar eclipse the Moon can come in front of the Sun to hide it completely as seen by us on Earth. When the Moon's apparent diameter is too small to produce a total eclipse a ring of light is still visible around the edge of the Sun. This is an annular eclipse. When the apparent diameter of the Moon is equal to or greater than the Sun, a total eclipse results observable when the shadow of the Earth passes over the lunar disc. The eclipse can be total or partial. In a lunar eclipse, the Earth comes directly between the Sun and the Moon.

EINSTEIN, *Albert* (1879-1955)

German mathematician and theoretical physicist whose work revolutionized modern science, especially astronomy. His special theory of relativity was put forward in 1905. He analyzed the concepts of mass and energy.

ELLIPSE

An ellipse is produced by projecting a circle at an angle. the importance of the ellipse in astronomy is seen in Kepler's first law. In fact, all planetary orbits are ellipses. In and ellipse, a line running from one "center" point to any place on the ellipse and back to a second "center" point is always the same length.

ENERGY (solar, stellar)

What is the source of energy that allows

stars to shine? the explanation came through the discovery of thermonuclear reactions. the initial transmutation of hydrogen into helium results in a very small loss of mass. this mass is turned into energy, and part of this energy is observed as starlight. Transmutation of other elements occurs later in a star's life.

EPHEMERIDES

Books and tables indicating the position of the planets, Sun and Moon among the stars and zodiacal constellations.

EUROPA

A satellite of Jupiter discovered in 1610.

EVENING STAR

The brightest "star" to be seen in the sky is the planet Venus. Apart from the Sun and Moon, it is the only heavenly body that is able to cast a shadow. Venus is also a morning star which is viewed in the sky in early a.m.

EXPANDING UNIVERSE

The universe of galaxies is expanding: the distances between galaxies continually increases. We have not yet seen to the "edge" of the universe.

FIELD OF FORCE

A solid body which has a magnetic core which is referred to in Newton's work as gravitational field.

FIRST DEGREE OF ARIES

The vernal equinox is the first degree of Aries.

FUNDAMENTAL ASTRONOMY

Fundamental astronomy was concerned with determining the positions of heavenly bodies, studying their motions, and discovering the theories governing these motions in order to predict for the future.

G.M.A.T.

This is the abbreviation of Greenwich

Mean Astronomical Time. Before January 1, 1925, all astronomical time was reckoned from zero hour beginning at mean noon: this avoided changing the date during the night. Greenwich Mean Time is solar time on the meridian of Greenwich, England.

Since January 1, 1925 G.M.T has been reckoned from midnight to midnight with the zero hour beginning from midnight to the following midnight.

GALAXY

It an assemblage of dust, gas, and billions of stars. The Milky Way is a galaxy. there are spiral galaxies, normal and barred. These are round shaped, with arms extended from the nucleus. Elliptical galaxies are spindle shaped with no arms. Irregulars are systems with no symmetry.

Edwin Hubble conceived the method of classifying galaxies that is still in use today. Our galaxy is a spiral-shaped galaxy.

GALILEI, *Galileo* (1584-1642)

He was an astronomer whose work initiated modern science. He left Pisa at the age of eleven to go to Florence, where his family originated. He returned to Pisa at age seventeen to study medicine. He was very critical and believed nothing unless he verified it himself. He was methodical and became a great physicist.

He invented a pendulum apparatus for timing pulse; watches were unknown at that time. He got into mathematics at age 25 in Pisa. He opposed some of the Ancients theories. It was during this period that he dropped a stone from the Leaning Tower of Pisa. This demonstrated the incorrectness of Aristole's view on falling bodies. Because of his innovative ideas he was received with hostility. So he again left Pisa and went to Florence, then served eighteen years as a professor at Padua. The turning point in his career was his use of his telescope for observation of the heavens. this telescope was constructed in 1609. He then discovered through his use of the telescope; 1) the nature of the Moon's surface, 2) the appearance of stars as points of reference, 3) the phases of Venus, 4) the four satellites of Jupiter and 5) sunspots. In 1615 he was denounced by the Catholic Church because of his teachings. He was forbidden to teach any theory that conflicted with scripture, especially the Copernican theory. He wrote much despite opposition from the church. In 1633 he was banished to an approved residence; it was there he completed another work that was secretly published in Amsterdam.

GANYMEDE

It is the fourth satellite of Jupiter and the third satellite Galileo discovered.

GAS

Gaseous matter is found everywhere in the universe. Cold bodies, such as planets, are usually surrounded by a gaseous atmosphere.

GEMINI - The Twins

The third constellation of the zodiac. Early identification was by the two chief stars named Castor and Pollux. It is near the celestial pole and can be seen in the morning sky from the end of summer. The Brighter star is Pollux.

GIANTS

Special type of star--yellow to red in color and their brightness is hundred times greater than normal stars like our sun.

GLOBULAR CLUSTER

Closely packed system of stars distributed in a globular spherical form. A system of stars that have a gravitational attraction to each other.

GRAVITY

Gravity is the force of the mutual attraction of matter and is responsible for the "fall" of objects to Earth.

HALLEY'S COMET

the most famous comet. Its appearance goes back to antiquity. It was first studied seriously by western scientists in 1680--when Isaac Newton formulated a method for calculating the parabolic orbit of comets.

HELIUM

A light, inert gas that is a major constituent of stars.

HEMISPHERE

One half of a sphere--as in northern, southern, eastern or western hemispheres.

HERCULES

A constellation seen in the summer evenings between Corona Borealis and Lyra. Hercules contains the spectacular M13 globular star cluster.

HERMES

An asteroid, or minor planet, Hermes was

discovered in 1937. Its period of revolution is 535 days.

HERSCHEL, *Sir William* (1738-1822)

Herschel was a German-born professional musician and amateur astronomer. He traveled to England in 1766. Too poor to afford a telescope, he decided to make his own (at that time they made mirrors from bronze), completing it is 1774. He made other as well. On March 13, 1781 with one of his telescopes he discovered what he thought was a comet; it turned out to be the new planet Uranus. He became famous immediately and received more than one medal. King George III appointed him court astronomer. He was no longer poor. For a while Uranus was called Herschel. He also discovered the sixth and seventh satellites of Saturn. Herschel did much double star research. His son, Sir John Frederick William Herschel (1792-1871) also a noted astronomer, continued his father's work.

HIDALGO

Minor planet, or asteroid. Hidalgo's orbit crosses the ecliptic (plane of Earth's orbit) at a great angle.

HUBBLE, *Edwin* (1889-1953)

This American astronomer is associated with the correct description of the universe outside the Milky Way Galaxy. He determined whether objects belonged to the Milky Way or were outside of it.

HUYGENS, *Christian* (1629-1695)

His ideas contributed to the development of Galileo's telescope. He was the first to introduce the use of a convergent eye-piece, the possibility of such was pointed out by Kepler.

HYDROGEN

Hydrogen is found everywhere in the universe. Hydrogen is present to a greater degree than any other element in the universe.

IAPETUS

A satellite of Saturn.

ICARUS

This minor planet, or asteroid, was discovered in 1949; it is one of the few that come within the orbit of Mars.

INFERIOR PLANET

A planet of our solar system whose orbit is within that of Earth. There are two of them: Mercury and Venus.

INTERSTELLAR MATTER

Matter that lies between the stars. Before the discovery of this matter between the stars scientists thought that interstellar space was a matter in various forms and sizes existing everywhere between the stars.

IO

One of the four Galilean satellites of Jupiter.

JANUS

A satellite of Saturn.

JUNO

A minor planet, or asteroid. Juno was the third asteroid to be discovered. The first two were Pallas and Ceres. Juno was discovered in 1834 by K. L. Harding.

JUPITER

The name Jupiter, the largest planet in our solar system, means "Lord of Heaven." Jupiter, along with Venus, is one of the brightest planets; it is easily visible with the naked eyed. The red spot on Jupiter was first seen by Cassini in 1665. Its color became apparent only when large telescopes were used.

KEPLER, *Johannes* (1571-1630) - See Part I

KEPLER'S LAWS

1) The orbit of a planet lies in an elliptical plane around the Sun--with the Sun at

one focus and nothing at the other focus.

2) A line between the Sun and a planet sweeps across equal areas of space in equal periods of time.

3) The cube of the radius of a planet's orbit about the Sun, measured in astronomical units equals the square of the period of its revolution, measured in years. One astronomical unit equals the distance from the Sun to the Earth.

LAW OF GRAVITATION

Two material bodies attract each other with a force which is directly proportional to their mass, and inversely proportional to the distance between them.

LEO

One of the constellation of the zodiac. Leo is the lion.

LEONIDS

One of the annual meteor swarms usually appearing at the end of the year in the head of the constellation Leo.

LIBRA

One of the constellations of the zodiac. Libra is the scales, the balance.

LIFE IN OTHER SOLAR SYSTEMS

Most scientists agree that to have life as we know it, the temperature, pressure, and composition of the atmosphere has to be conducive to that of life. The planet probably has to be a certain distance from the Sun. If the planet is too close it is too hot, life would burn up. If the planet is too far away we would freeze. Chances are, only one planet per solar system could sustain life as we know it.

LIGHT YEAR

One light year equals 9.5 trillion kilometers: the distance traveled by light in one year at 300,000 km/sec. It is the usual way to express distance in the universe.

LUNA

Poetic name given to the Moon. Also the name of a Russian spacecraft sent to the Moon.

LUNAR CRATER

A cup-shaped depression formed by the impact of a meteorite.

LUNAR ECLIPSE - See Part I

LUNATION

Sequence of phases of the Moon, starting with New Moon and ending with Balsamic Moon. See Astrology section for greater descriptions.

LYOT, *Bernard* (1897-1952)

Solar astronomy completely changed as a result of his invention of the coronagraph. It enables astronomers to study the Sun's corona at any time and location, not just for a few minutes during a solar eclipse at a certain locality.

LYRA

A summer constellation whose main star is Vega.

LYRIDS

Annual swarm of meteorites which can be seen between April 20 and April 23.

M

This letter stands for stellar classification, and is an abbreviation of Messier. Charles Messier (1730-1817) was a French astronomer who assigned numbers to stellar objects such as star clusters and nebulae. these numbers are still used today. Messier's catalog included 107 objects (including tow duplicates).

MAGELLANIC CLOUDS

One of the grand sights of the southern sky. They are the nearest galaxies to ours. Because of their closeness to our Milky Way Galaxy we have been able to study their makeup--something that is impossible to do with Andromeda which is in the northern sky farther away. The large "cloud" is 160,000 light years away; the small one, 190, 000.

MAGNITUDE

The apparent brightness of a star or planet. The unaided eye magnitude of stars or planet rages from +5 (dim) to -27 (bright). The Moon is considered magnitude -14 and the Sun is magnitude -27.

MARS

One of the planets of our solar system, easily recognized by its red color. Mars magnitude ranges from -1 to -2.8. It is the next planet out from the Sun after Earth. Of all the planets, Mars is the best for observation. There has been much theory about possible inhabitants at one time or another. Some theorists believe life existed but was dried out. The diameter of Mars is about half that of Earth.

MASS

The amount of molecular material in a given object. Mass causes an object to have weight in a gravitational field. The greater the mass the greater the weight. The weight of a person if transported to the Moon would be 1/6 of the weight as on earth. If we could stand being on the Sun, our weight would be 28 times greater. However, even the weight of a person on Earth varies according to the latitude and altitude he is located.

MERCURY

Mercury is one of the two planets whose orbits are within that of Earth. It is the closest planet to the Sun. The orbit of Mercury has the second greatest eccentricity of any of the major planets in our solar system. Pluto's orbit is the most eccentric.

METEOR

A flash of light in the sky caused by a

fragment of metal or rock as it plunges through Earth's atmosphere. Some people call them shooting stars. Fragments are meteoroids while in space but meteorites when they strike Earth. The larger meteorites can hit the Earth intact when they fall, or they can break up into small pieces. Scientists say some craters are definitely from meteorites as evidenced by nearby fragments of meteoric origin. They consist mainly of iron and nickel. Smaller meteoroids disintegrate before they hit the ground.

MILKY WAY

The name for our galaxy. According to Greek mythology Hercules was placed at the breast of Juno while asleep, some of her milk fell into the heavens where it spread through that part of the sky to be called Milky Way. And if you find it in the sky one might think of spilled milk or perhaps scattered diamonds.

MIMAS

The first satellite of Saturn, the second nearest the planet was discovered by William Herschel on September 17, 1789. Its period of rotation is 22 hours, 37 minutes and 5.3 seconds. Like all other satellites of Saturn, it is very close to the plane of the ring.

MINOR PLANETS

Small bodies in the solar system which have been discovered since the nineteenth century are also called asteroids. The larger asteroids are Ceres, Juno, Pallas, and Vesta. There are thousands of smaller ones. Most orbit the Sun between Jupiter and Mars. Some are rocky, consisting mostly of silicate material; others are metallic, largely nickel and iron.

MIRANDA

The fifth largest satellite of Uranus was discovered by Gerald P. Kuiper on February 16, 1948. Its period of rotation is one day, 9 hours, 36 minutes.

MIRROR

Flat mirrors are used in reflecting and refracting telescopes to redirect the path of light. Concave mirrors are used to magnify images in reflecting telescopes.

MOON

Our moon is the only natural satellite of our Earth. Even though our Earth has only one satellite, that satellite is one of the largest moons in our solar system. It is exceeded in size by the two chief satellites of Jupiter, Ganymede and Callisto, the first satellite of Saturn, Titan, and the first satellite of Neptune, Triton. The next two satellites of Jupiter, Io and Europa are a little larger and a little smaller, respectively, than our moon.

MOONLIGHT

Light of the Sun reflected off the surface of the Moon.

MORNING STAR

This is another name for Venus. When that planet is visible in the east just before sunrise.

NADIR

That point on the celestial sphere which is diametrically opposite the zenith. It is vertical when projected downwards.

NEPTUNE

The planet Neptune's atmosphere like that of Uranus, contains methane, as well as hydrogen, helium and ammonia. The cloud-top temperature is about -216 degrees centigrade. The first of Neptune's satellites was discovered by W. Lassel very shortly after the discovery of the planet itself in 1846. The satellite was named Triton. It revolves around the planet in 5 days, 21 hours, 2 minutes, 38.04 seconds. It is one of the satellites in the solar system that has a retrograde motion. In 1949, Kuoper discovered the second satellite, Nereid, which has a most unusual orbit for a satellite. The orbit is very elongate. The period of revolution is almost one terrestrial year: 359 days, 21 hours, 9 minutes.

NEWTON, *Isaac* (1642-1727)

Born on January 4, 1642. He explained that the ocean's tides are caused by the gravitational pull of the Sun and the Moon. Newton also cleared up the mystery of the comets, showing that their motions obeyed the same law as the planets. Newton formulated three laws of motion:

1) A body at rest or in uniform motion in a straight line continues that way unless acted upon by an outside force.

2) The acceleration of body is inversely proportional to its mass and directly proportional to the net force acting upon it. Acceleration is also in the same direction as the force.

3) To every action there exists an oppo-

site and equal reaction.

He was a physicist and a very capable mathematician. He discovered the complex nature of white light, and split it into the seven colors of the spectrum. Newton was fascinated with the study of light. He also became interested in metaphysical consideration, chronology and theology. He talked about the Apocalypse and the prophecies of Daniel. He was modest, and sought tranquility. When it came to his work, he was very vocal and determined. He is buried at Westminster Abbey where a monument marks his grave.

NIGHT

This is the period between sunset and sunrise. During the winter, polar night lasts six months.

NOVA

A star which exhibits a sudden brightness, increasing rapidly to a maximum in several day. The outburst is an eruption of an old, white dwarf star caused by transfer of matter from a nearby normal star in the sky where no star was seen before.

OBERON

This is the second largest satellite of Uranus and most distant of the large moons.

OCCULTATION

Disappearance of one celestial body behind another in the course of its motion.

OPHIUCHUS

A large constellation to the north of Scorpio. Some think it should be part of the zodiac. Barnard's Star, named after astronomer E. E. Barnard, is a red dwarf in Ophiuchus.

OPPOSITION

When a planetary body is directly opposite the Sun, as seen from Earth.

ORBIT

Closed curve or path that one celestial body makes around another. Most orbits are elliptical.

ORION

It is the easiest constellation to recognize. It describes a giant who is armed with a club. It appears as a rectangle with the giant's belt through the center. The main brilliant stars are Betelgeuse which has a red cast and Rigel which is blue-white.

OXYGEN

This element composes nearly a third of the Earth by weight. It is the most abundant element in the Earth's crust, and is essential for aerobic respiration and life.

OZONE

Ozone is a gas made up of molecules

containing three atoms of oxygen each. As an upper atmospheric gas it absorbs ultra-violet radiation.

PALLAS

Pallas is a minor planet, or asteroid. It was discovered March 28, 1802 by W. Olbers (1758-1840). It orbits between Mars and Jupiter. Pallas moves around the Sun in 1684 days. Along with Ceres, Juno, and Vesta it is one of the four largest minor planets.

PEGASUS

A large constellation in the northern hemisphere. It extends over more than 1100 square degrees and contains the square formed by Pegasus and Andromeda.

PERIGEE

The point in the orbit of a satellite where it is nearest to the center of the Earth.

PERIHELION

The point in the orbit of a planet where it is nearest to the Sun.

PERSEIDS

This is the best known meteor shower, appearing as "shooting stars" around the middle of August in and near the constellation of Perseus.

PHASES

Several planets and satellites show phases. The Moon, illuminated by the Sun in its orbit around the Earth, shows phases from new (dark) to crescent, quarter, gibbous, full, and back again. The inferior planets Mercury and Venus, which come between the Sun and the Earth, also exhibit phases.

PHOBOS

The first satellite of Mars, discovered on August 17, 1877 by Asaph Hall, the same day as the second satellite, Deimos.

PHOEBE

Ninth satellite of Saturn, discovered August 16, 1898 by Pickering (1846-1919).

PHOTOSPHERE

The outer layer of the Sun which provides the light and is the visible surface of the Sun. It is the source of the visible radiation which we receive from the Sun.

PIONEER

This is the name given a series of spacecrafts launched by the United States into interplanetary space. The earliest one was to the Moon. Pioneer 1 and Pioneer 2 were launched in the first part of 1958. Pioneer 3 failed at launching. Pioneer 4 was launched March 3, 1959 and ended up being in planetary orbit, becoming Earth's second planetoid.

PISCES

A zodiacal constellation whose symbol is the fishes.

PLANET

A planet is a body which revolves around the Sun. Earth is one of them. Other planets can be seen in the night sky because they reflect light from the Sun. Five of them can be seen with the naked eye: Mercury, Venus, Mars, Jupiter, Saturn. They have been known since ancient times because of their distinct motion against the background of the constellations.

Mercury, because it is close to the sun, is low in the sky and visible before sunrise or after sunset.

Venus, also close to the Sun, is the brightest body in the sky during practically the whole time that it is visible. During six months of the year it is a morning star; for the other six months it is an evening star.

Mars is the red planet. It can be recognized by its reddish tint.

Jupiter, the largest planet in the solar

system, is almost as bright as Venus.

Saturn is also know for its brightness. It has a yellow tint. Saturn, when seen through a telescope usually shows its spectacular rings.

The planetary orbits are in this order. Mercury, Venus, Earth, Mars, Jupiter, Saturn, Uranus, Neptune, Pluto.

PLANETARIUM

A building where the motion of the celestial bodies can be projected on a domed ceiling.

PLANETOID

An older name for minor planet, or asteroid.

PLEIADES

A small blue-white star cluster in the constellation Taurus. At least six can be seen with the unaided eye, with dozens more visible through a large telescope. The seven brightest, known as "Seven

Sisters," have been traditionally associated with Spring and agriculture.

PLUTO

The discovery of Neptune was based on perturbations in the motion of Uranus, for which Neptune was responsible. Similarly, perturbations of Neptune's orbit led to the discovery of Pluto, our smallest major planet, by Clyde Tombaugh in 1930.

POLARIS

This star is about one degree from the celestial north pole.

POLE STAR, North

At this time, the north celestial pole is in the constellation Ursae Minoris (Little Bear). Plolaris, the pole star is about 50 arc minutes from the pole. Due to the precession of the Earth's axis, the north pole star has changed since the Egyptian Pyramids were built.

POLLUX

The companion of the star Castor in the zodiacal constellation Gemini. Pollux is a reddish yellow star.

PRECESSION OF THE EQUINOXES

The Earth's axis of rotation describes a cone about the pole of the ecliptic. At the same time the equator remains perpendicular to it and thus the intersection of this plane with the ecliptic moves retrograde. It takes 25,800 years to complete the precession of the equinoxes back to the original position. We are now entering the Aquarian Age. Many astronomers take this into account when they give positions in the ephemerides. Some think that this gives a different position in the year for the zodiacal signs as follows:

Capricorn - 1/19-2/17

Aquarius - 2/17-3/13

Pisces - 3/13-4/19

Aries - 4/19-5/15

Taurus - 5/14-6/21

Gemini - 6/21-7/21

Cancer - 7/7-8/11

Leo - 8/12-9/18

Virgo - 9/19-11/1

Libra - 11/1-11/23

Scorpio - 11/23-11/30

Ophiuchus - 11/30-12/18

Sagittarius - 12/19-1/19

PROGRESSED SIGNS

(including Ophiuchus)

Aries - 4/19-5/15

Taurus - 5/14-6/21

Gemini - 6/21-7/21

Cancer - 7/7-8/11

Leo - 8/12-9/17

Virgo - 9/21-11/1

Libra - 11/1-11/23

Scorpio - 11/23-11/30

Ophiuchus - 11/30-12/18

Sagittarius - 12/19-1/19

Capricorn - 1/19-2/17

Aquarius - 2/17-3/13

Pisces - 3/13-4/19

PROGRESSED SIGNS

(without Ophiuchus)

Capricorn - 1/19-2/17

Aquarius - 2/17-3/13

Pisces - 3/13-4/19

Aries - 4/19-5/15

Taurus - 5/14-6/21

Gemini - 6/21-7/21

Cancer - 7/7-8/11

Leo - 8/12-9/17

Virgo - 9/18-11/1

Libra - 11/1-11/23

Scorpio - 11/23-12/19

Sagittarius - 12/19-1/19

PROPER MOTION

Actual motion of the stars-not due to rotation of the Earth. Barnard's star moves 1 degree every 360 years. the Sun moves 12 miles per second.

PROXIMA CENTAURI

A red dwarf star in the triple Alpha Centauri System. Currently the closest star to Earth. Also known as Rigil Kentaurus. Visible from the southern hemisphere.

PULSATING UNIVERSE

This is another name for oscillating, or expanding universe.

QUADRANT

One of four equal parts; the arc of the circle that is 90 degrees.

QUASAR

Short for quasistellar object. A powerful source of energy more distant than most of the visible galaxies and possessing an energy greater than that of any galaxy.

RADAR

Acronym for "radio detection and ranging." Directional radio waves sent out from an antenna,strike an object and bounce back as an echo. It is received by a ray tube which produces an image.

RADIO ASTRONOMY

Study of electromagnetic radiation received from the sky in the wavelength between 30 m and 1 mm. Developed between 1932 and 1940, radio astronomy is currently in a state of active development and will greatly add to our knowledge of the universe.

RADIO TELESCOPE

Apparatus in the shape of a large reflector designed to catch radio emissions from the sky.

RANGER

A name given to a series of space probes launched towards the Moon. Ranger 1, launched August 23, 1961 failed to leave orbit. Ranger 2, November 18, 1961: disappeared on the day of launch. Ranger 3, January 26, 1962: missed the moon and went into orbit as a planetoid. Ranger 4, April 23, 1962: crashed on the far side of the Moon. It failed to transmit any photographs. Ranger 5 became another planetoid. Ranger 6, January 30, 1964: the first of the series to follow exactly the right course. After a slight correction when half way to the Moon, it crashed at the programmed position. But again, no photographs. Ranger 7, July 28, 1964: brought the final success which made up for all the earlier setbacks. The course was correct; so was the place it landed. More than 4000 pictures were taken and sent back to Earth under great conditions.

Because of these pictures it became possible to reconstruct the terrain of the Moon.

Ranger 8, February 17, 1965: reached the Moon and took thousands of pictures. Ranger 9, March 21, 1965: was the last of the series. It landed where it was sent.

RED SPOT

A large spot in the south zone of Jupiter's disc. At some times it is particularly distinct and is a deep rose. It was discovered by Cassini in 1665. It is a large oval-shaped hurricane.

REFLECTOR TELESCOPES

This instrument has a concave mirror and an eyepiece to collect light, magnify celestial images, and bring them to focus for observation. The mirror performs the same purpose as the glass of a refractor. There are various types of reflecting telescopes: Newtonian, Schmidt Camera, Casserain, Maksutov. The largest singlemirror reflector is the 236-inch telescope in the Special Astrophysical Observatory near Cherkassk, Russia.

REFRACTOR TELESCOPE

This instrument collects light from a celestial object with a lens instead of a concave mirror.

RETROGRADE MOTION

The motion of all planets takes place in a direct, west to east direction. However, when planets catch and pass each other in the sky they sometimes seem to move east to west in a retrograde motion.

RHEA

Second largest satellite of Saturn.

Diameter is 1530 km. The period of revolution is 4 days, 12 hours, 25 minutes. Rhea was discovered by Cassin at Paris Observatory on December 23, 1672.

RIGEL

This star in the constellation Orion, is one of the brightest in the sky. It is a white star. This star is a binary system: Rigel and a companion star revolve around a common center of gravity. Rigel is 815 light years away from us.

RINGS OF SATURN

A remarkable ring system encircles Saturn. It appears to be a thin flat ring, but research has shown the rings are not solid. They are composed of countless tiny particles--mostly ice. The seven major rings range in thickness from about 650 feet to nearly 10,000 feet.

ROTATION

The Earth rotates, or spins, on its axis once every 23 hours, 56 minutes. The Sun rotates in about 25 days. Other planets, and our whole Milky Way Galaxy also rotate.

SAGITTARIUS

This is a constellation of the zodiac and is identified as the archer.

SATELLITE

This is an object that revolves around a planet. It is smaller than the planet. It is called a satellite because it is smaller and is held in orbit by the large sphere. You can tell a lot about a planet including mass by the actions of the satellite of the planetary body.

SATELLITE, Artificial

Since 1957 the Earth has acquired a number of very small artificial satellites which have been launched by man into long-term orbits. In 1966 there were about 500 of them. These satellites carry instruments which can send information back to Earth by radio contact.

SATURN

This planet is the marvel of the solar

system because of its famous ring system. Until the discovery of Uranus Saturn was the most distant planet from antiquity. Saturn is a dull yellow color, with a leaden hue. Lead has been long associated with Saturn. Venus and Jupiter appear to be silvery and Mars decidedly red. Saturn is very far from the Sun. The surface temperature is in the region of -150 degrees centigrade. The temperature would be even lower if Saturn didn't have a thick layer of clouds of methane and ammonia. these cover the planet and retain some of its heat. The satellite of Saturn called Titan is the only one that is easy to see. Iapetus was discovered in 1671. Rhea, 1672, Tethys and Dione, 1684. Mimas and Enceladus were discovered in 1789. They were followed by Hyperion, 1848, Phoebe, 1898, and Themis, 1900. Another satellite was discovered in 1967 which is called Janus.

SCORPIO

This is a constellation of the zodiac and

its symbol is the Scorpion.

SEASONS

The four divisions of the year of a planet whose axis of rotation is not normal to the plane of its orbit. The sun illuminates the two hemispheres with unequal intensity because one is tilted away from the Sun in summer and winter. The result is four seasons as we are familiar with one Earth. These seasons are fixed by the equinoxes and the solstices.

SHADOW

Shadows seen on celestial objects often provide and impression of reality and contrast. On the lunar surface shadows show the shape as well as the breaks in mountain chains, such as the Apennines. Shadows also show the isolated peak of Pico, etc.

Phases of Venus and Mercury and the Earth's Moon are the result of planet-wide shadow. Because of their orientation to Earth and of the Sun they sometimes appear to be only partially illuminated .

SHOOTING STARS

Most people call meteors shooting stars.

SIDEREAL

Sidereal time is based on the rotation of Earth with respect to the stars.

SIDEREAL PERIOD

The period of rotation of a planet in sidereal time (for Earth: 23 hours, 56 minutes, 4 seconds).

SIRIUS

the brightest star in the sky. Sirius is in Canis Majoris, a southern constellation. It is a white star. Sirius was observed visually in 1862 by an amateur astronomer Alvan Clark.

SOLAR CYCLE

the mean period of about eleven years

which coincides with sunspot activity.

SOLAR PROMINENCES

Solar prominences are jets of gas thrown outward from the Sun.

SOLAR SYSTEM

This is our immediate world . . . one Sun with planets revolving around it.

The planets in order around the Sun are Mercury, Venus, Earth, Mars, Jupiter, Saturn, Uranus, Neptune, Pluto. Some small planets or asteroids lie between Mars and Jupiter. The four largest were discovered in 1801, 1802, 1804, 1807: Ceres, Pallas, Juno, and Vesta, respectively.

A. Charlois (1864-1910) of Nice, France was a well-known asteroid "hunter." He has 101 to his credit. At this time there are more than 3000 asteroids.

SOLAR WIND

Stream of atoms and ions emitted by the

Sun and thrown into space at high speed. Earth's magnetic field causes these particles to interact with atoms in the upper atmosphere and yield the aurora borealis and aurora australis light.

SOLSTICE

The time when the Sun reaches its northern (summer) or southern (winter) point farthest from the celestial equator. the summer and winter solstices occur, respectively, about June 22, and about December 22.

SPECTRUM

Sequence of colors produced from light by using a prism.

SPIRAL GALAXY

Many galaxies have remarkable shapes. Some consist of two arms in the form of a spiral, starting from a central nucleus. Our Milky Way galaxy is of this type. About eighty percent of all the known galaxies are of this type. Some galaxies are barrels, which means the spiral arms do not start from the nucleus but from the ends of two straight arms on opposite sides. Others are ellipsoidal or irregular.

SPUTNIK

Russian word meaning satellite and a name given the first series of Soviet launched satellites.

STAR

A heavenly body that shines with its own light produced by thermonuclear fusion.

Variable stars are stars whose brightness varies with time.

Novas are stars that exhibit sudden bursts of brightness. Novas are white dwarfs pulling gas away from larger companion stars. The brightness is caused by the fusion of the incoming matter.

Double stars or binaries . . . are stars that come in pairs. The stars revolve around a common center of gravity.

White dwarf star, main sequence, giant, supergiant: these are the categories into which most stars are placed. The white dwarf stars are very small and faint but hot. The giants and supergiants are large, with low temperatures. Nevertheless, because of their size, they are quite luminous.

STAR CLASSIFICATIONS, Major

O The hottest star . . . temperature around 40,000 K (Kelvin)

B Hot stars . . . 20,000 K

A White stars . . . 10,000 K

F Yellowish stars . . . 7,500 K

G Yellow stars . . . 5,500 K

K Reddish stars . . . 4,500 K

M Red stars . . . 3,000 K

SUMMER

On Earth the season between summer

solstice and the autumnal equinox. Summer is the warm season because the Sun is higher in the sky. Thus, its rays shine on the ground more directly and intensely. The longest day comes at the beginning of the Summer season.

SUN

The sun is a star which has the Earth as one of its captured planets. It is a large gaseous ball of hydrogen (75%) and helium (25%). The pressure and temperature increase with depth to the center. The central temperature is 15 million K. The Sun's diameter is 1,390,000 km.

SUNSPOT

The Sun is not so pure a fire as was originally thought. One of the earliest discoveries made due to the invention of the telescope in 1610 was the presence of many dark spots. sunspots are "cooler" areas on the Sun that contain intense magnetic fields. Changes in sunspot activity seems to come in eleven year cycles.

At the beginning of a cycle spots start to appear at middle latitudes (about 35 degrees). They later appear near the equator. Galileo was the first astronomer to observe sunspots.

SUPERIOR PLANET

A superior planet is one whose orbit lies farther from the Sun than the Earth's orbit: Mars, Jupiter, Saturn, Uranus, Neptune and Pluto.

SUPERNOVA

An exploding star--caused by transfer of matter to a white dwarf or by collapse of a massive star.

TAURUS

This is a zodiacal sign whose symbol is the bull.

TELESCOPE

This is an instrument used for gathering light and magnifying distant objects.

TETHYS

Third satellite of Saturn, discovered by Cassini, March 21, 1684.

TIDE

Movement of ocean water is effected by the combined gravitational influence of the Sun and the Moon. the gravitational attraction of the Moon is much stronger than that of the Sun even though its mass is less than the Sun. This is because it is closer, High tides occur approximately once every 12 1/2 hours.

TIME

There are three forms of time.

Mean time: the hour angle of the Sun.

Local time: when the mean Sun comes to the meridian at any particular place.

Civil time: also universal time, with 0 at midnight.

There are twenty five time zones, most are limited by 15 degrees wide. Sidereal time is the hour angle of the first point of Aries.

Albert Einstein was the first scientist to show that time is relative, and that it cannot be separated from the concept of space.

TIME ZONES

It has been convenient to divide the Earth's longitude into 15 degree zones. The meridian of Greenwich, England, is the starting point for time zones.

TITAN

The largest satellite of Saturn. It was discovered March 25, 1655.

TITANIA

This is the largest satellite of Uranus (1,610 km diameter). It was discovered by Herschel on January 11, 1787.

TRITON

this is the largest satellite of Neptune

(about 3,400 km diameter). It was discovered October 10, 1846, by Lassell (1799-1880).

TROJANS

This is a group of minor planets. Orbits for more than 3,000 minor planets have been calculated most between the orbits of Mars and Jupiter.

TROPICAL PERIOD

The period of the Earth's revolution around the Sun. This determines the return of the seasons.

UMBRIEL

This is the third largest satellite of Uranus (about 1,200 km diameter). It was discovered by Lassell on October 24, 1851--the same day the first satellite, Ariel, was discovered.

UNIVERSAL ATTRACTION

In 1543 Copernicus stated the concept of

the Sun as a central star (or power) with a procession of planets in orbit around it.

In the early 17th century Kepler published three famous laws.

1. The orbit of a planet lies in an elliptical plane around the Sun--with the Sun at one focus and nothing at the other focus.

2. The relation between the radius of a planet's orbit about the Sun, measured in astronomical units (A) and the period of its revolution, measured in (T) is T squared = A cubed. (One astronomical unit equals the distance from the sun to the earth.)

3. A line between the Sun and a planet sweeps across equal areas of space in equal periods of time.

UNIVERSE

This is a term which has a more precise meaning in astronomy than everyday language conveys. It means all that has ever been observed.

URANUS

This was the first planet to be discovered since remote antiquity. It was accidently discovered by William Herschel on March 13, 1781, when he was observing through his 16 cm telescope.

VEGA

Star in the constellation Lyra--one of the brightest in the northern sky.

VENUS

Outside of the Sun and Moon, this planet can be brightest in the sky. It can be more brilliant than any star. It is also known as the morning star, evening star, and shepherd's lamp. The phases of Venus are similar to the phases of the Moon.

VERNAL EQUINOX

First point of Aries, and the beginning of spring when days and nights are equal length.

VESTA

A minor planet (or asteroid). Discovered at the turn of the nineteenth century.

VIRGO

Constellation on the Zodiac. Symbol is the virgin.

VOID, Interplanetary

At one time scientists believed much of space was a vacuum. It is now known this is not true. Gas and dust particles are widely scattered.

ZODIAC

The signs or constellations located along the ecliptic.

BIBLIOGRAPHY

Chartrand, Mark R. *Skyguide*. New York: Golden Press, 1982.

Hall, Manly P. *Astrological Keywords, Sixth Edition*. Los Angeles: The Philosophical Research Society, 1958.

Muller, Paul. *Concise Encyclopedia of Astronomy*. Chicago: Follett Publishing Co., 1968.

Seeds, Michael. *Foundations of Astronomy, Second Edition*. Belmont: Wadsworth Publishing Co., 1988.

Sun Bear and Wabun. *The Mecicine Wheel*. New York: Prentice Hall Press, 1980.